Patrick Moore's Practical Astronomy Series

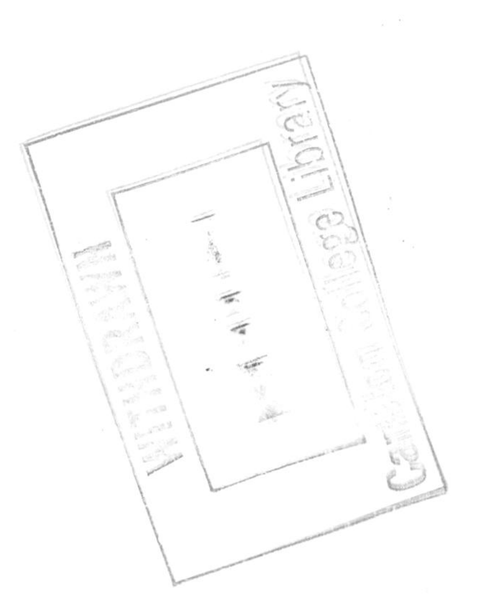

Other Titles in This Series

Telescopes and Techniques (2nd Edn.)
Chris Kitchin

The Art and Science of CCD Astronomy
David Ratledge (Ed.)

The Observer's Year (Second Edition)
Patrick Moore

Seeing Stars
Chris Kitchin and Robert W. Forrest

Photo-guide to the Constellations
Chris Kitchin

The Sun is Eclipse
Michael Maunder and Patric Moore

Software and Data for Practical Astronomers
David Ratledge

Amateur Telescope Making
Stephen F. Tonkin (Ed.)

Observing Meteors, Comets, Supernovae and other Transient Phenomena
Neil Bone

Astronomical Equipment for Amateurs
Martin Mobberley

Transit: When Planets Cross the Sun
Michael Maunder and Patrick Moore

Practical Astrophotography
Jeffrey R. Charles

Observing the Moon
Peter T. Wlasuk

The Deep-Sky Observer's Year
Grant Privett and Paul Parsons

AstroFAQs
Stephen Tonkin

Deep-Sky Observing
Steven R. Coe

Field Guide to the Deep Sky Objects
Mike Inglis

Choosing and Using a Schmidt-Cassegrain Telescope
Rod Mollise

Astronomy with Small Telescopes
Stephen F. Tonkin (Ed.)

Solar Observing Techniques
Chris Kitchin

Observing the Planets
Peters T. Wlasuk

Light Pollution
Bob Mizon

Using the Meade ETX
Mike Weasner

Practical Amateur Spectroscopy
Stephen F.Tonkin (Ed.)

More Small Astronomical Observatories
Patrick Moore (Ed.)

Observer's Guide to Stellar Evolution
Mike Inglis

How to Observe the Sun Safely
Lee Macdonald

The Practical Astronomer's Deep-Sky Companion
Jess K. Gilmour

Observing Comets
Nick James and Gerald North

Observing Variable Stars
Gerry A. Good

Visual Astronomy in the Suburbs
Antony Cooke

Astronomy of the Milky Way: The Observer's Guide to the Northern and Southern Milky Way (2 volumes)
Mike Inglis

The NexStar User's Guide
Michael W. Swanson

Observing Binary and Double Stars
Bob Argyle (Ed.)

Navigating the Night Sky
Guilherme de Almeida

The New Amateur Astronomer
Martin Mobberley

Care of Astronomical Telescopes and Accessories
M. Barlow Pepin

Astronomy with a Home Computer
Neale Monks

Visual Astronomy Under Dark Skies
Antony Cooke

Creating and Enhancing Digital Astro Images

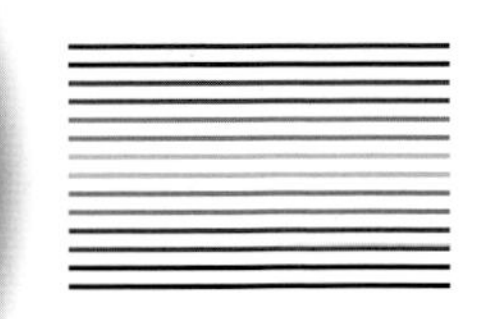

Grant Privett

Grant Privett
Ledbury
Herefordshire
United Kingdom
e-mail: g.privett@virgin.net

British Library Cataloguing in Publication Data
A catalogue record for this book is available from the British Library

Library of Congress Control Number: 2006928730

Patrick Moore's Practical Astronomy Series ISSN 1617-7185
ISBN-10:1-84628-580-1
ISBN-13:978-1-84628-580-6

Printed on acid-free paper.

9 8 7 6 5 4 3 2 1

Springer Science + Business Media
springer.com

What one fool can do, another can.
Ancient proverb

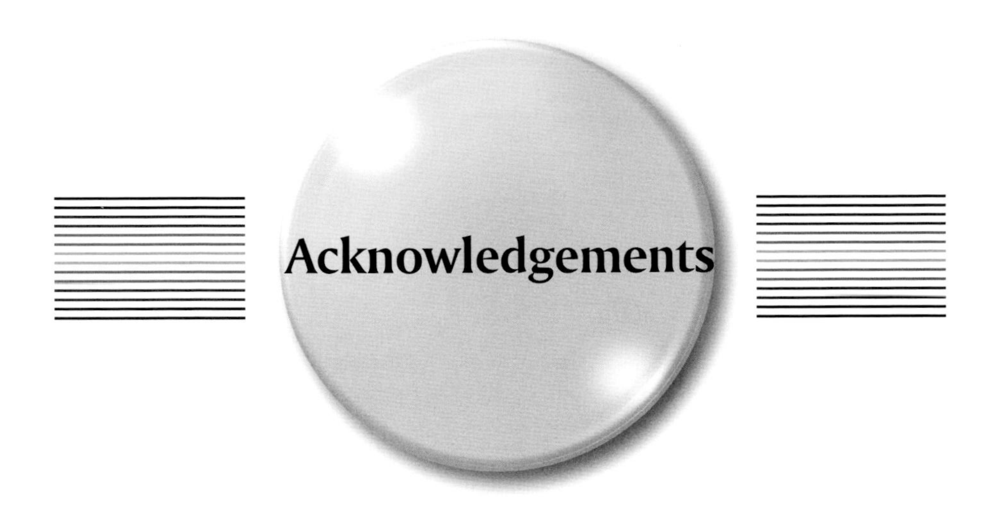

Acknowledgements

Once more, I owe a great debt to Rachel without whose unfailing support this book would never have appeared. As before, she put up with the late nights, my interminable moans about the clouds and the fact that the DIY just didn't get done. It will now, I promise.

In addition, I would like to thank the many astro-imagers who so generously allowed me to use their images to illustrate the examples of image processing herein. I owe each of you an inordinately large beer. Anyone seeking details of the equipment and techniques they use to take such fine images need look no further than Chap. 15: "Notes on Image Contributors", where you will find the URLs for some spectacularly illustrated websites.

I would like to extend my thanks and gratitude to David Woodward, Kev Wildgoose, Mark Wiggin and John Watson who were brave enough to read the early drafts, identify omissions and provide many useful suggestions for improvements.

Contents

1 **The Joy of Image Processing** 1

2 **A Digital Image – The Basics** 3

3 **Choosing Your Camera** 9

4 **Acquiring Images** 19

5 **What Is Best for ...** 27
The Moon and Planets 27
Deep Sky 29
Aurora and Meteors 30
The Sun 31
Comets 32

6 **Displaying Images** 35

7 **Image Reduction** 37
The Importance of Dark Frames 37
Taking Your Dark Frames 40
Creating Master Darks 41
Flat-Fields 44
Processing Your Flat-Field 47
Bias Frames 48
Image Transformations 50
Image Stacking 51
Image Mosaicing 56

Automated Image Reduction 61
Image Handling 63
Histograms 65
Brightness Scaling 67
False Color Rendition 71

8 Other Techniques 75
Statistics 75
FITS Editor 76
Pixel Editing 77
Bloom and Fault Correction 77
Night Visualization 79
Profiles 79
Isophotes 79
Polar Representation 79
Bas Relief and 3D 81
Blinking 82
DDP 84
Edge Detectors 84
Masks and Blending 86

9 Image Enhancement 87
Background Gradient Removal 88
Image Sharpening 91
Image Smoothing 92
Median Filtering 93
Unsharp Masking 95
Image Deconvolution 97

10 Handling Color Images 101
Acquiring Color Composites 101
Processing Color Composites 104
Achieving Color Balance 106
Narrow-Band Filter Composite Images 108

11 Handling Image Sequences 113
Image Acquisition 113
Image Quality Techniques 115

12 Astrometry and Photometry 121

13 The Problems 127

14 Postscript 129

15 Notes on Image Contributors 131

16 Appendix **133**
Software 133
Hardware 137
Further Reading 138
Acronyms and Abbreviations 139

Index **141**

CHAPTER ONE

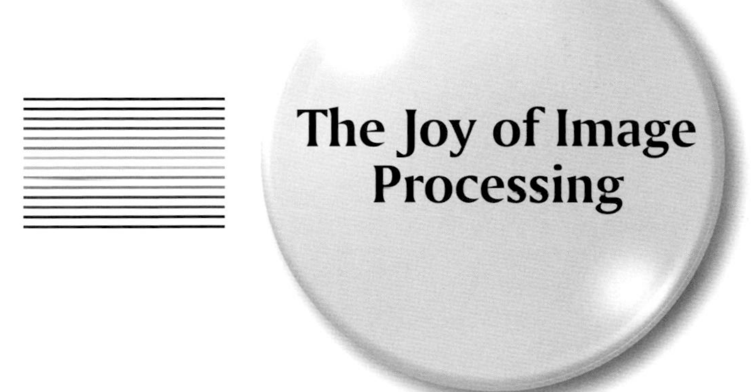

The Joy of Image Processing

So, what is image processing? Well, at its most basic, it's the gentle art of manipulating a digital image to show its content to the best possible or desired effect.

Obviously, the effect you require from your images will vary from night to night. On one night you may be trying to determine the magnitude of the faintest star you can image, on another the emphasis may be on picking out delicate areas of nebulosity around the Pleiades, while on still another, you might be striving to record detail on a Jovian moon.

All are examples of astronomical imaging, yet each requires the use of a distinctly different approach; both to the imaging process and the subsequent image processing you apply. This is to some extent true for normal day-to-day photography, but where astronomical image processing differs from the handling of holiday snapshots is that some essential processing stages *must* be undertaken for every image. Whether you take your images with a purely scientific or aesthetic objective in mind, these crucial processes must be performed for the best possible image to be created. For the subsequent processing stages, the techniques used are frequently a matter of choice or taste and the rules about their application more arbitrary. The essential stages are referred to as "image/data reduction" while the later stages are generally described as "image enhancement".

In many ways the image reduction process is relatively straightforward and routine, and it is mainly the diverse range of image enhancement possibilities that leads to the tendency for amateur astronomers to reprocess their previously taken images as they become more practiced and their expertise grows. The practice of returning to old images is often quite educational and also has the virtue of making even a cloudy night bearable.

This book will lead you through both stages of image processing while explaining the concepts involved through clearly illustrated examples rather than through a mass of bewildering equations. For those of you keen to go deeper into the mathematical details lurking tantalizingly behind your computer's graphical user interface (GUI), a number of references are identified as sources of useful further reading.

The equipment essential for undertaking image processing are a computer, some image processing software and a digital camera of some form; be it a charge-coupled device (CCD), webcam or digital single-lens reflex camera (DSLR). To make selecting a sensor a little easier, the early chapters discuss their relative merits and limitations. No sensor is perfect for every task and there is always an element of horses-for-courses involved. Observers who have previously been active astro-photographers using film and slides can create digital images from their slides, prints or negatives using scanners. They will also be delighted to find that work at a PC may bring new life to old images.

On the subject of software this book is deliberately non-committal as many programs exist to do image processing and image reduction, and, frankly, none are perfect and those that approach it are clearly overpriced at present. Consequently, the book normally discusses the principles involved rather than the detail of any given software. The book does provide a list of useful and popular software that is well worth a look.

The other requirement – the most important thing – is a willingness to learn, and the endurance to keep on nagging away at a problem until it is sorted out. Some of the concepts in image processing can be a little confusing, but with familiarity the confusion gradually wears off and things start to make more sense.

Figure 1.1. An SBIG ST7-MXE CCD camera. In this case it is shown fitted with a standard camera lens. The ST7 camera has been around for some years, but upgrades to the chip sensitivity and the connection type to the computer – in this case USB – mean this remains a popular camera. Image credit: Santa Barbara Instruments Group.

CHAPTER TWO

A Digital Image – The Basics

Well, what is a digital image? Perhaps it is easiest to start with one of the most familiar types of image we encounter. A printed photograph – whether the picture originates from a digital or film-based camera – will consist of a rectangular area covered with tiny dots of varying intensity and color. When viewed from a distance the dots merge and the print shows how the photographed scene would have appeared had you been there to see it. The fact that a photograph is made up of dots – known as pixels – is not normally apparent unless you try to enlarge it too much or use a magnifying glass.

The crystals that make up a film emulsion-based photograph – in effect defining the pixels – are packed together closely, but are randomly located within each frame, whereas digital images employ a grid-like layout made up of small rectangular (or square) light-sensitive pixels. The variation in brightness or intensity of these pixels represents a two-dimensional (2D) view of the world – or, in the case of astronomy, space. Unsurprisingly, digital images are like this because most cameras employ sensors that consist of a flat rectangular patch of silicon that has been divided up into a grid of minute light-sensitive rectangles.

The sensors have electronics associated with them that can quickly determine how much light has impacted onto each of the light-sensitive pixels during the exposure. These sensors are placed at the focal plane – i.e. the point where the light comes to a focus – of a camera lens or telescope and, as a result, a digital image is generated by determining a numeric value indicating how much light fell onto each pixel of the image. The numeric value for a given pixel will not be the same as the number of photons collected but will certainly be closely related and broadly proportional to it. Creating this gridded list of numbers is quite an undertaking, given that quite inexpensive digital cameras can now boast 6 million pixels, while professional photographers often use cameras containing 15 million pixels.

Figure 2.1. A CCD sensor; the light-sensitive silicon chip that forms the basis of many CCD cameras. This example is the KAF-0402E that appears in the popular SBIG ST7-ME camera. Image credit: Santa Barbara Instruments Group.

So, we have a regular grid of millions of light-sensitive pixels. That's a good start, but there are other important aspects of the image to be considered, such as whether the image is color or monochrome, what the spatial resolution is and how many shades of color (or gray, in the case of a monochrome image) the resultant image contains.

Imagine if you took an image where the full Moon just fitted into the sensor. To determine the resolution – the ability to record fine detail – of the image you would divide the width of the Moon into arc seconds (which is, conveniently, roughly 1/2 degree or 1800 arc seconds) by the number of pixels across the image. For many modern digital cameras this might give a result of the order of 1–2 arc seconds per pixel. However, it's best to determine the field of view of a sensor before buying it, and fortunately, there is a simple method related to the physical size of the pixels. You will find that, typically, pixel sizes vary from as small as 4 microns (a micron is one-thousandth of a millimeter wide) to around 25 microns. The pixel size will normally be found on the website of the camera manufacturer.

The way to estimate the resolution of a telescope/sensor combination is to do the calculation that follows. Don't worry, this isn't as bad as it sounds – all the trigonometry has been done for you in creating the number 206 – see below.

1. Multiply the size of your pixel (using microns) by 206.
2. Divide the number you obtained in stage 1 by the aperture of the telescope or camera lens (using millimeters).
3. Divide the number you obtained in stage 2 by the focal ratio of your telescope.

Not painful, and it yields an answer measured in arc seconds per pixel.

For those of you who like formulas, this is equivalent to:

$$\text{Resolution} = \frac{\text{Pixel size} \times 206}{\text{Focal ratio} \times \text{Aperture}}$$

where pixel size is expressed in microns and the focal length is expressed in millimeters. Note that a focal ratio has no units, it's just a number. If you do a calculation and get silly results check that you have got the other units right. Using a focal length expressed in feet or pixel size expressed in inches or furlongs is not going to work. Note: 1 inch = 25.4 mm.

If the value is not that required for a given purpose, an optical device called a Barlow lens can be employed to increase it, or another optical device called a focal reducer used to decrease it.

As an example; consider a telescope of 200 mm aperture with a focal ratio f/6.3 used with a camera boasting 10 micron pixels:

Stage 1. $10 \times 206 = 2060$
Stage 2. $2060 / 6.3 = 326.9$
Stage 3. $326.9 / 200 = 1.63$.

So the resolution of each pixel is 1.63 arc seconds.

For those uncomfortable with math, Tables 2.1 and 2.2 provide some typical values for popular telescope and lens systems.

Table 2.1. The sky coverage – resolution – of each pixel when using telescopes

Aperture (mm)	Focal ratio (f/)	Resolution 7.5 micron pixel	Resolution 10 micron pixel	Resolution 20 micron pixel
60	15	1.71	2.28	4.56
114	8	1.69	2.25	4.51
150	8	1.28	1.71	3.41
200	4.8	1.61	2.15	4.29
200	6.3	1.23	1.64	3.27
200	8	0.96	1.28	2.56
200	10	0.77	1.03	2.05
250	4.8	1.29	1.72	3.44
250	6.3	0.98	1.31	2.61
250	10	0.62	0.83	1.65
300	6.3	0.81	1.08	2.16
300	10	0.51	0.68	1.36

Table 2.2. The sky coverage – resolution – of each pixel when using camera lenses

Focal length (mm)	Resolution 7.5 micron pixel	Resolution 10 micron pixel	Resolution 20 micron pixel
28	55.2	73.6	147
35	44.1	58.8	118
70	22.1	29.4	58.9
135	11.4	15.2	30.4
200	7.72	10.3	20.6
300	5.15	6.86	13.7

As we now know the size of each pixel in our sensor in terms of arc seconds, we can readily derive the region of sky covered by our image. It will be the resolution of each pixel multiplied by the number of pixels across the sensor's width. As before, let's take it in stages.

1. Multiply the resolution of sensor pixel (using arc seconds) by the number of pixels along the axis.
2. Divide the number you got from stage 1 by 3,600.

As a formula this is:

$$\text{Coverage} = \frac{\text{Pixel Resolution} \times \text{Number of pixels}}{3{,}600}$$

So, as an example, if we consider the LX200 mentioned earlier, our pixel size is 1.63 arc seconds and the width of our image is 1200 pixels:

Stage 1. $1.63 \times 1200 = 1956$
Stage 2. $1956 / 3600 = 0.543$;

then the sky coverage amounts to a bit more than half a degree.

Now the catch is that if you wanted pictures with even better resolution – as you would for detailed lunar or planetary imaging – you could use a Barlow lens which, in effect, magnifies the image that appears at the focal plane, or you would employ a sensor with smaller pixels. So a pixel size of 1 arc second per pixel would become 0.5 arc seconds with a ×2 Barlow lens and 0.33, 0.25, 0.2 arc seconds with a ×3, ×4, ×5 Barlow lenses, respectively. Similarly, focal reducers can be used to reduce the effective magnification of the scope. These are generally more limited in their effect upon magnification, with a factor of ×0.5 or ×0.33 being typical. With a ×0.5 reducer in place, a 1 arc second pixel would become a 2 arc second pixel.

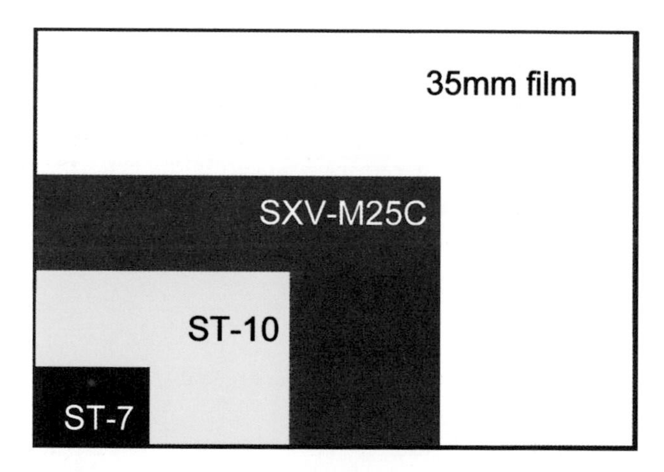

Figure 2.2. The relative sizes of the sensors found in some well-known cameras; the SBIG ST-7, the SBIG ST-10 and the Starlight Xpress SXV-M25C. These are compared against the size of film emulsion. Some sensors are now as big as 35 mm film, but they are still rather expensive. Image credit: Grant Privett.

Be sure to bear in mind the fact that the resolution values we have calculated represent only the theoretical resolution of the sensor pixels with a given telescope/optics combination. The actual resolution achieved may be constrained by the resolving power of your telescope, and just as importantly, the steadiness of the atmosphere you are observing through. It is a rare night indeed in the UK or much of the US that telescopes get to perform to their full potential – it appears to be pretty much reserved for island mountains, Antartica and the high plateaus of Chile. On a turbulent night your pixels will still have the same angular resolution spacing but all sharp edges or features within the image will be blurred by the atmosphere.

As examples, a 75 mm (3 inch) aperture will be able to distinguish between two equally bright stars approximately 1.5 arc seconds apart, while a 200 mm (8 inch) aperture will do the same with stars 0.6 arc seconds across. Unfortunately, in urban areas the air turbulence will reduce this to a varying figure of the order of 2–3 arc seconds.

Image sampling is another area worth considering. If you are imaging the planets or are simply imaging star fields you will find that the most aesthetically pleasing results are obtained only when the image is over-sampled. That is to say you are operating with a camera that has a higher resolution than the finest feature you wish or hope to observe. For example, if you are looking at a double star whose components are separated by 3 arc seconds you would just be able to resolve it with an imaging sensor/telescopes combination where the pixels are of 3 arc second resolution, but the image would look pretty awful. A much nicer result would be obtained if the camera employed had a resolution of 1 arc second or less. In the 3 arc second resolution image the stars will be very blocky and angular, at best showing the double star as two bright pixels adjacent to each other, with most stars represented by a single pixel. With below 1 arc second resolution the component of the double star will appear as minute circles or crosses of light with a dark space clearly separating the double star. The down side of this is that if you use the same telescope for both images the higher resolution image would be rather dimmer, as the light from the star must now be spread out over 9 (3×3) times as many pixels. Obviously to overcome this you can take longer exposures or stack images – but more of this later. The upside, and it is considerable, is that over-sampling allows stacked planetary images to clearly show tiny features such as Encke's division on Saturn – a very narrow gap in the rings that eludes most visual observers. Thierry Legault and some others have imaged features on the moons of Jupiter (diameter < 2 arc seconds) when working at a focal ratio of more than 40.

As a general rule choose a pixel size, Barlow lens/focal reducer, telescope combination that has a resolution that is at least twice as fine as the smallest feature you hope to image. This reasonably intuitive rule, more formally known as the Nyquist limit, can be derived from some rather lengthy mathematics involving Fourier transforms which the reader will find discussed in the standard texts on image processing mentioned in the references section. Obviously it can be taken too far. If you over-sample too much, the light will be so spread out that it's difficult to tell if there is anything in the picture at all.

Another defining characteristic of an image is the extent to which it resolves differences in intensity. If we take the monochrome image case first, we find that it is usually expressed by describing an image as of 8, 12, 14 or 16 bits resolution. For those unfamiliar with "bits" they refer to the way numbers are stored in computer memory. The larger the number of bits employed in storing a number, the larger the number that can be stored. In the case of imagery it often refers to the number of gray levels that the camera can detect between a totally black scene – when no light has been received – and a perfectly white scene – when the sensor has recorded as much light as it can cope with. Table 2.3 shows the most common values.

These days monochrome CCD cameras generally operate at 16 bits, with a few still using 12 or 14 bits and some older auto-guider cameras employing 8 bits. Some cameras at professional observatories now use 24 bits per pixel (or more) but those are not generally available on the commercial market. It should be noted that 14 bits represents the minimum number of bits required to allow sophisticated image enhancement techniques to be undertaken without artifacts becoming apparent in the resultant image. You can work with 12 bit imagery, but it will limit what you can hope to get from the results.

However, the situation is different – and slightly confusing – for color cameras such as webcams, DSLRs and some CCDs, in that the number of bits the image is recorded in may be quoted as 24, 36 or 48 bits. This is because each pixel of the output image will have been generated using information from three closely spaced pixels on the sensor; one each of red, green and blue. So for each pixel of a 36 bit camera, the intensity of the red light sampled will have been determined to 12 bit accuracy as will the values for the other colors. Adding together these three sets of 12 bits makes up the 36 bits mentioned. The number defines the ability of the camera to resolve differences in the total light intensity, while the ratios between the values found for each pixel will define the hue of color recorded. Obviously, it might be expected that the best possible results would only be obtained using 36 or 48 bits per pixel images that permit a wider range of pixel brightness values. Yet, when multiple images are stacked, this potential limitation is dramatically overcome as the plethora of exquisite planetary images taken with 24 bit cameras so graphically testify. It is perhaps significant that the well-known planetary imagers Damien Peach and Martin Mobberley have recently moved over to using the Lumenera LU 75 cameras (12 bit) from a Philips ToUCam to improve their pictures still further.

Table 2.3. Shades of gray resolvable in images of various "bits" of accuracy

Bits	Discernible shades of gray
8	256
10	1024
12	4096
14	16384
16	65536

CHAPTER THREE

Choosing Your Camera

The most important things you need to decide when selecting a camera is what you want to use it for and how much money you have available to spend. After considering that, you really should ask a further question: "Am I sure?", before proceeding. For many people, the answer is going to be that they want to image *everything*, in which case the camera selected must inevitably be a compromise based upon the other equipment – telescopes and lenses for example – available.

Many people are attracted to astronomical imaging by the pictures published in popular astronomy magazines such as *Astronomy Now* and *Sky & Telescope*. These usually break down into two camps, deep sky images showing extended regions of hydrogen gas and dust and, at the other extreme, sharp crisp color images of Mars, Jupiter or Saturn hanging against a pitch black background. Unfortunately, these two specializations have very different hardware requirements.

In the deep sky example what we need is sky coverage which, as we have seen, translates to a big chip used with a telescope or a smaller chip used with a camera lens. At present (Spring 2006) cameras boasting a light-sensitive area 25 mm (1 inch) or more across remain rather expensive – it can be safely assumed that this will gradually change with time, but camera prices in the last 10 years have certainly not dropped as fast as the chip price would suggest. The price of an entry-level camera seems quite stable, but what you get for the money steadily improves – hopefully cameras will become genuinely affordable to all astronomers in time.

Currently large chips are most commonly found in DSLRs and some cooled color cameras such as the Santa Barbara Instruments Group's (SBIG) ST-2000XCM or Starlight Xpress SXV-M25C. They have some problems in that, being made up of millions of pixels, the chips can be slow to read out. Consequently, you should look for a camera that is compatible with the widely used USB computer interface

Figure 3.1. The business end of a Starlight Xpress SXV-M25C CCD camera. Generally, the sensor size of affordable CCDs isn't yet the same as film emulsion, but in this case the sensor is certainly big enough to image a large chunk of sky. Image credit: Starlight Xpress.

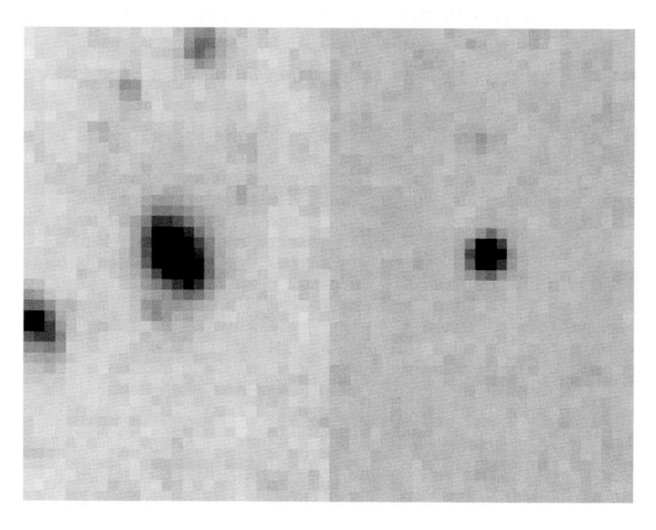

Figure 3.2. Star images being distorted by optical flaws. In this case on- and off-axis images taken with a zoom camera lens. The off-axis region was also severely vignetted. Image credit: Grant Privett.

which permits a fast readout – certainly a lot faster than the parallel port interface it has taken over from. Image download/read times for USB are usually less than 10 seconds even for the bigger cameras. Currently the standard to look for is USB 2 rather than the older and slower – but still pretty quick – USB 1.1 that preceded it.

Another problem encountered with very large chips is that they can be so wide that they occupy an area on the focal plane that is bigger than the "sweet spot" of the optical system used. This means that toward the edges of the image you will see evidence of distortions with star images displaying coma – the star images are distorted into V-shaped fans with the V pointing toward the center of the field of view. You will also find evidence of vignetting where the amount of light reaching the most off-axis regions of the focal plane is diminished and the image corners appear considerably dimmer. Both these problems can be resolved, the former by the use of a coma corrector or different telescope, and the latter by a technique called flat-fielding – neither problem need be a complete show stopper.

At the other end of the spectrum are the planetary observers who are, instead, trying to observe targets that may be only a few arc seconds across. As examples, Mercury and Saturn are typically a mere 6 and 17 arc seconds across respectively. For the planetary observer the instrument of choice is a telescope with a large focal ratio – normally referred to as the f/number – representing the ratio of the focal length to the telescope aperture. Many observers swear by Newtonian reflectors that have a focal ratio of 8 or 10, because even a simple ×2 Barlow lens may be enough to ensure good sampling. Unfortunately, we must note that any telescope more than a meter long can move around in a breeze, making imaging difficult. So it is hardly surprising that many observers compromise on the rather squat telescope tubes associated with Schmidt–Cassegrain (SCT) and Maksutov telescopes. Their relative portability is also a big plus.

It goes without saying that for any imaging the telescope should have the best quality optics you can sensibly afford, and a stable mount. This should be coupled to a drive that can track accurately for at least 30 seconds – otherwise you are going to have to take an awful lot of pictures and combine them later. It's fair to say that there's an

Figure 3.3. An example of image vignetting showing the pronounced darkening at the edges and corners of the image. In this example the fall off from full illumination is quite drastic. In other instances it may be a lot more subtle. In all but the most severe cases flat-fielding can help make the whole image usable. Image credit: Gordon Rogers.

Figure 3.4. An example of an internal reflection generated within the optics of a cheap SLR camera. Note how the brightest part of the image – the prominences – form a displaced ring. The contrast has been exaggerated to make it more easily visible. Image credit: Grant Privett.

ongoing argument regarding whether SCT and Maksutov telescopes are suitable for planetary imaging of the highest quality, due to the size of the aperture taken up by a large secondary mirror. While it is true that refractors do not suffer from this problem, they certainly do suffer from achromatic aberration – light of different colors coming to focus at very slightly different places – and so they have failings of their own. The argument between refractor and SCT also tends to ignore the fact that some of the best planetary imagers of the last 10 years have been taken using SCTs.

Whilst the ideal solution for an observer trying to image galaxies and planets is to have two telescopes, this is not an option open to all of us. The best compromise instrument for all-round imaging is arguably an f/6 Newtonian reflector, as this has a fairly small secondary mirror, is easy to collimate and does not suffer from achromatic aberration – it also has the advantage of being considerably cheaper than alternatives of similar aperture. If deep sky is your primary interest then an f/4.5 instrument becomes more interesting, and if planetary imaging is your passion an f/8 system has clear advantages, with a Barlow lens employed to raise that to somewhere in the range f/20–f/30 when imaging planets.

Alternatively, some people use their camera with a telescope to image planets, but then use it in conjunction with a camera lens – preferably mirror-based – to image the deep sky. Often, this is accomplished by the camera riding piggy-back upon the telescope with the telescope mount providing the ability to track the stars. This can work superbly, especially when it is used in conjunction with the filters to compensate for the poor focus achieved by most achromatic camera lenses at infrared (IR) wavelengths.

If you already have a telescope, things are rather more difficult, as the room for manoeuvre is greatly reduced, but typically you should aim for 0.3 arc second resolution on the camera when imaging the planets and roughly between 1 and 2 arc seconds when imaging large deep sky objects.

Returning to the choice of camera, there is an important choice to be made regarding the use of color. Some cameras such as the SAC, Starlight MX7C, SBIG ST-2000XCM and most webcams or DSLRs are "one-shot" color cameras. As might be expected, these allow you to take a single image which contains all the color and intensity information you need to form a picture. They do this via an ingenious modification to the camera sensor. Instead of each pixel being sensitive to light of any color between IR and ultraviolet (UV), minute on-chip filters are employed so that some pixels are only sensitive to red light while the others are sensitive to either green or blue light. To avoid IR light messing up the color balance during everyday use, the on-chips filters also reject IR light.

This sort of camera is wonderful for quickly taking a color picture, and the many great pictures taken using them make it obvious how successful they can be. However, the on-chip filter does make it rather difficult to obtain high-quality monochrome images of very faint objects. This is because a proportion of the light reaching each pixel is being discarded – absorbed – by the on-chip filters, so fewer photons reach the light-sensitive pixels. Throwing away the IR certainly does nothing to help, as many stars and deep sky objects emit quite a lot of light at wavelengths our eyes do not see – the UV is mainly absorbed by water in the atmosphere. This lack of sensitivity can be something of a limiting factor, but if you have a telescope that can provide well-guided exposures of 30 minutes or more this is not a great issue.

For those chasing faint objects, such as new supernovae or previously unknown asteroids, it is probably best to buy a monochrome CCD camera and to use it together with a filter wheel or slide when color imaging is needed. A filter wheel is a device that, when rotated or slid, places a filter in the light path to the CCD. This

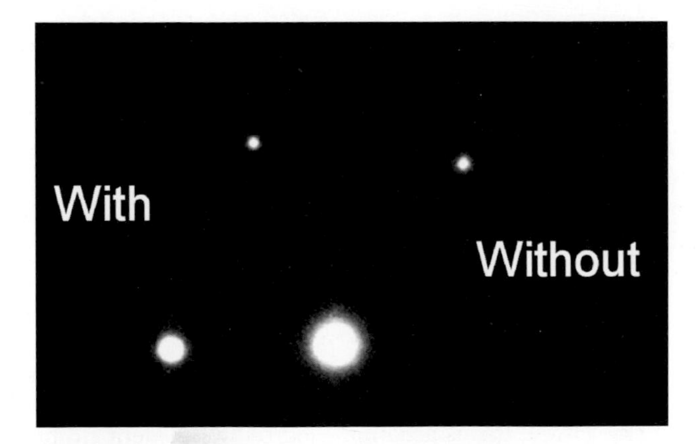

Figure 3.5. Two monochrome images of Alberio taken through an achromatic refractor, showing images of the double star with and without an infrared filter. The infrared component is not as well focused and so the star image bloats when the filter is not used. The bluer fainter star is also affected but not as much. Picture credit: Grant Privett.

allows you to achieve the maximum sensitivity of your system by taking an unfiltered image. This can be surprisingly faint – 20th magnitude objects are accessible from semi-urban sites using telescopes of 200 mm aperture and 23rd magnitude has been attained by some very diligent observers willing to spend several nights on the same object.

On those occasions when you instead wish to use the camera for color purposes you just take three images of the object; one using a red filter from the filter wheel, one a green filter and the last with a blue filter. These images are then combined, to yield a colored image that is generally better than you would obtain using a one-shot color camera! Created from images captured using red, green and blue light filters, the output image is said to be RGB true color. An alternative approach is to use cyan, magenta, yellow filters thereby forming a CMY image.

Both of these RGB and CMY color imaging approaches can be augmented by including an unfiltered image. This means that the images are created using the unfiltered image as luminance – most easily thought of as approximating brightness/intensity – with the RGB or CMY images providing the color of each pixel. The best of both worlds.

Unfortunately, this can be trickier than you might expect, as you have to ensure the focus is correct for each filter and that, when combined, the images are very accurately aligned to each other. This can all become a bit tedious when imaging planets – hence the popularity of one-shot color cameras for planetary work and the rise of webcam-specific software such as *RegiStax* and *AstroStack* which automatically handle the alignment, dark subtraction, flat-fielding and summing.

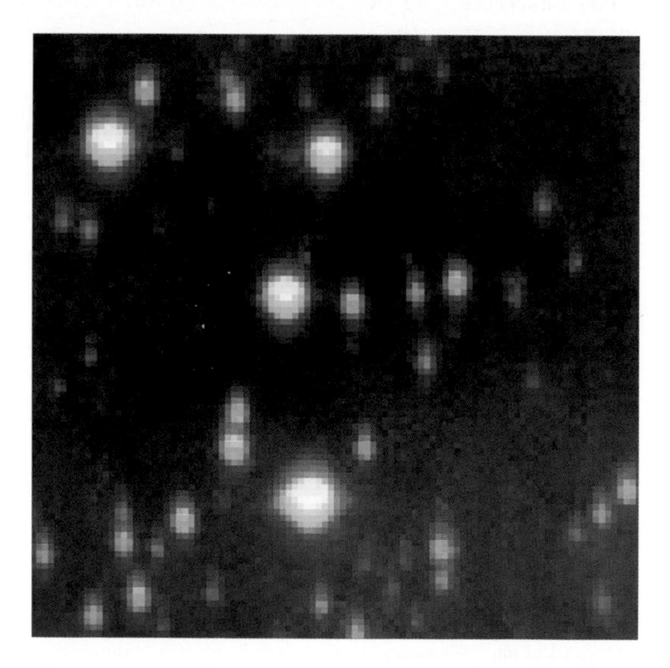

Figure 3.6. When creating a color image from individual red, green and blue images, the alignment of the images is critical. If they do not align, false color edges and images will abound. Image credit: Grant Privett.

To summarize; perhaps it's easiest to consider the disadvantages of each camera type.

DSLRs usually have quite large sensor multimillion pixel chips and so are well suited to deep sky imaging. However, they are based around uncooled one-shot color detectors so you may need to be prepared to use quite long exposures – it's worth checking that the one you buy can do long exposure, as not all do. They are not very sensitive when imaging diffuse emission nebulas, as the IR filters they employ usually cut out some of the visible deep red light received.

DSLRs do have the enormous advantage that they can also be used for family snapshots and "normal" photography. It is essential that any DSLR specified must be operated in its highest resolution raw data mode – most have one – despite the fact that size of the images created will rather eat up the card memory. Good examples of less expensive, but capable cameras are the Canon EOS 350D "Digital Rebel XT", the Canon EOS 20D (the 20Da is even better but more expensive) and the Nikon D70 and Pentax *istD.

Each is being superceded by a version with even more pixels and more sophisticated functions, but at the time of writing their prices (with lenses) are currently rather steep, i.e. >£900 (including lenses) in the UK, but around $700 in the US. Be careful to ensure that you buy a DSLR from which the lens can be removed. Without it most astro-imaging is not possible.

Webcams usually contain small chips with small pixels that make them ideally suited to planetary observing. The fact they can be read out very quickly is a distinct advantage as it increases the chances of "freezing" the seeing during a

Figure 3.7. The Nikon D70s DSLR camera. Simple to use and providing high-quality results. Image credit: Nikon Inc.

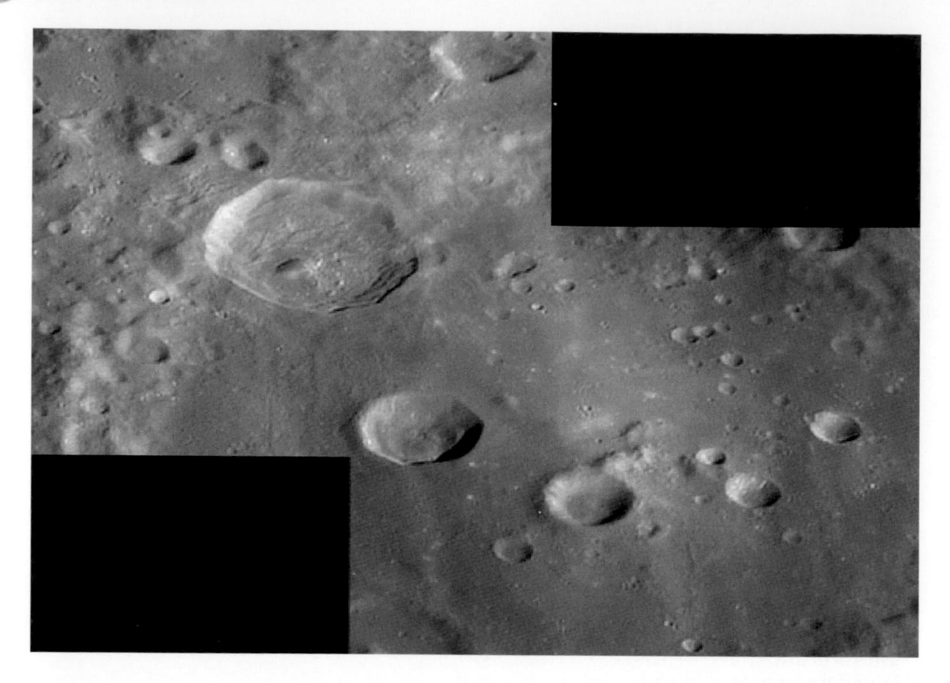

Figure 3.8. An image of the Moon taken at f/40 using a Celestron 356 mm (14 inch) SCT. The field of view is entirely within the walls of the crater Clavius, with the largest crater visible the 54 km wide Rutherford. The smallest features resolved are a mere 600 m across. Created from a stack of 900 images, this gives some idea of the breathtaking potential of webcams used with commercial telescopes. Image credit: Damian Peach.

moment when it was perfect and allows you to take a considerable number of frames before the rotation of the planet becomes evident.

Good examples are the Philips ToUCam Pro and modified webcams supplied by SAC or ATIK. The unmodified webcams are very affordable items – some comparable in price to a medium-range Plossl eyepiece or a pair of budget 10×50 binoculars.

CCDs can contain small or large chips and so can be employed for planetary or deep sky observing. The trend in recent years has been for webcams to take over as the planetary imager of choice, with CCDs repositioned in the market as the best choice for imaging very faint objects, or color imaging where a fast frame rate is not required. The popularity of one-shot cameras is growing fast, but at present most CCDs are monochrome, which means that in addition to normal imaging they are ideally suited to the pursuits of astrometry or photometry, where having as many photons as possible is essential.

It is worth remembering that some types of CCD camera support auto-guiding. The SBIG ST7-MXE – and many others they sell – operate by having two CCD sensors mounted in the camera head. The larger chip images the target, while the smaller adjacent chip stares at a star just off the main field of view. By reading out the smaller chip every second or so and looking at the position of the star images,

Figure 3.9. The business end of an SBIG ST-2000 CCD camera showing the main sensor chip and the smaller guiding sensor chip next to it.
Image credit: Santa Barbara Instruments Group.

it can detect drive errors which it then corrects by sending a stream of corrective signals to the drive controller via the auto-guider port.

Many of SBIG's competitors use two entirely separate camera heads for the same purpose: one for imaging via the focusing mount, and the other looking through the finder scope at a guide star or the target itself. The difference is that the smaller sensor images are not incorporated with the main imaging sensor in the camera head, which means more leads are needed – and there is more to go wrong.

An alternative to these is provided by the MX series of cameras from Starlight Xpress. These do very clever things with the chip readout and use a single chip to image and guide simultaneously – though it does mean that to get a 15 minute exposure you have to wait 30 minutes. Not an ideal solution, but very elegant in concept and leading to cameras as small and light as the SBIG solution.

All three solutions work very well, but you will need to carefully consider your budget and equipment before deciding which will best suit you. Certainly good guiding will make a huge difference to your success rate by greatly reducing the number of poorly guided images, and increasing the quality of those you do get.

An increasingly important consideration is whether you wish to control the camera remotely. This has become quite popular with people using GOTO telescopes where the observer is situated in the warm indoor environment (with toaster and coffee pot close at hand – bliss on a cold night) with the telescope perhaps in the garden or in extreme cases even on a different continent. Some cameras, notably the Apogee family, now contain network cards making the whole process much simpler. Alternatively, it is possible to run a low-spec computer in the observatory close to the telescope and use the *pcAnywhere* software to operate it remotely.

As you can see, the main types of camera are really quite different and best suited to different purposes. The ideal solution is to have one of each. Unfortunately, this is not possible for most of us. The simplest realistic rule is that if you are imaging planets use a webcam, and if chasing deep sky objects try a CCD or, failing that, a DSLR. At the time of writing, uncooled CCDs are coming onto the market – most famously as the SAC, Meade DSI and ATIK cameras – and these are blurring some of the distinctions. They are making the hobby much more accessible, financially, and can produce quite impressive results. To make life easier, the Meade camera can even be used to image deep sky objects using an alt-azimuth mount, if the exposures are kept to less than 60 seconds or so. Such cameras represent a good entry-level buy.

CHAPTER FOUR

Acquiring Images

The overwhelming urge when you first get your camera is to take some images – any images will do. This is hardly surprising as, after spending the cash, everyone is anxious to know whether or not they have wasted their money. However, taking good images is not a skill learnt in a few minutes. It takes quite a while to become proficient at taking images that are going to really reward the hours that can be spent at a computer reducing a disk full of raw data down to a single image you might want framed on the wall. There are lots of things to be done before you get close to saving an image, and most of them have to do with the telescope itself. To make things easier, I shall present them as a series of numbered paragraphs, each of which are addressed carefully by the best imagers. Some of them you may not be able to implement due to equipment limitations or other constraining circumstances. Those of you using camera lenses instead of telescopes will also find some of the information below of interest.

1. **Mount stability:** Ensure your telescope is stably mounted, and protected as much as possible from buffeting winds. The stability is essential, as a mount that moves during the course of a set of exposures will lose its polar alignment, and objects will gradually drift out of the field of view, thus necessitating regular adjustments to the declination axis. This can be a particular concern if you are observing at a site where no tarmac or concrete is available as the base. Watching the accumulating effects of a tripod sinking slowly into damp soil over the course of a night can be very frustrating. Try not to walk round too much as the vibration may well travel through the ground.
2. **Drive accuracy:** The other important constraint on the type of mount you should buy is that it provides accurate tracking of the object you are looking at. Simply stated, what you probably need as a starting point is a German equatorial mount – common with Newtonian tube assemblies – or fork

mounted telescopes such as a Schmidt–Cassegrain or Maksutov fitted with an equatorial wedge. If you are interested primarily in imaging the Sun, Moon or planets, then the short exposures involved mean a relatively poor drive quality is still usable. Even an alt-azimuth mount such as that found in the Meade ETX EC is usable, but if you intend to use exposures of more than 10–20 seconds duration you will need to be a bit more demanding. The best-known, affordable, yet good quality equatorial mounts are supplied by Vixen, Losmandy and Takahashi.

Good quality fork mounts are made by Celestron and Meade. There are many other very good manufacturers such as Software Bisque, who make the superb – but rather expensive – Paramount, who are also worth a look if sufficient funds are available. There are other "players" in the field at the more affordable end of the market including Synta who make the interesting EQ6 Pro mount – but those named earlier have a long-term reputation for consistent quality. It is wise to keep an eye on telescope/mount reviews in magazines for up to date information.

The aim should be to have a drive that has a "periodic error" of less than 20 arc seconds (smoothly varying, preferably) and then, if affordable, exploit any auto-guider capability provided by your camera. That should help to eliminate the impact of the drive error, and allow you to take longer exposures without obvious trailing of the image.

3. **Focus:** The focus of the instrument you use must be perfect. Not nearly, or pretty good, but perfect. Many observers find that it takes longer to focus an

Figure 4.1. Two faint star images; one showing trailing caused by drive error and the other with near perfect drive accuracy. Image credit: Grant Privett.

Figure 4.2. The renowned Celestron C14 SCT optical tube assembly mounted upon a CGE mount. A fine OTA for planetary imaging. Image credit: David Hinds.

object than it does to star hop to it and then center it. Focus really cannot be rushed and the expression "That's good enough" is the wrong mindset. If, early on in your imaging career, you get used to the fact it can take 20 minutes to focus the telescope, you will avoid having to revisit objects later to get pictures with tight star images. It took me some time to learn this, and I wish

I had done so earlier – lots of wasted clear nights. Tools are available to help with focus. These include aperture masks and occulting strips that cross the field of view thereby distorting the image of the star and making it easier to see when focus has been achieved.

Similarly, you can buy focusers with electronic control, fine adjustment or, even, larger knobs for better adjustment. These all help make it easier to spot whether a star is in focus, and are especially useful on telescopes with a focal ratio below f/6 where the difference between well-focused and nearly-focused will be less than a tenth of a millimeter (100 microns). Other ways to judge focus include looking at the number of faint stars in the field of view, or measuring the highest pixel count in a star from image to image. Be prepared to refocus several times during the night if the air temperature is changing. It can be a lot of work – even with motorized focus units from the likes of JMI – and not really what you want to do, but the effort will be repaid. A poor focus will ruin a good image.

Focusing is most easily determined by examining the image on a monitor with the image being regularly updated. It can be quite a slow and difficult process if you use a DSLR and have to examine each image through a view finder.

4. **Seeing conditions:** Only image seriously on nights of good seeing – use the others for practice. Turbulence in the atmosphere can swell a star image so that it is horribly blurred and becomes a distended blob. The best seeing is most frequently encountered only after midnight – when the ground is cooling and most home central heating has long turned off. The first sign of a good night is that when you look at bright stars they are only twinkling occasionally – obviously the sky should also be as transparent as possible for deep sky work. For planetary observers a faint haze or mist can be a surprisingly good thing as it is often associated with favorably still conditions in the upper

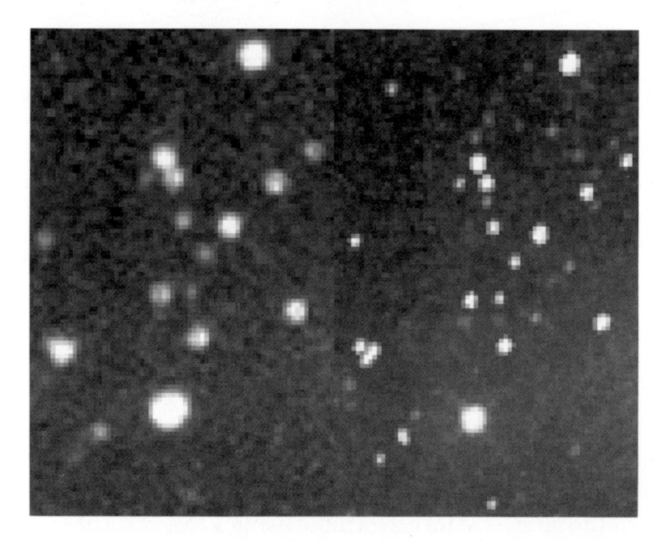

Figure 4.3. Two images showing an image taken on a night of poor seeing with a "that will do" focus (left) and a night of good seeing (right). Image credit: Grant Privett and Ron Arbour.

Figure 4.4. The galaxy NGC 6946 on the Cepheus/Cygnus border. The image shows the supernova of 2004 as imaged with a home-made 406 mm (16 inch) Newtonian reflector employing a friction drive. Note the careful attention that has been paid to achieving a good focus. Image credit: Ron Arbour.

atmosphere. You will probably find that your region will have its own climatic quirks that provide a hint. For example, when the jet stream across the Atlantic goes north of the UK the country generally gets better conditions. Similarly observers in other parts of the world will have their own special conditions. In an ideal world we would all have the sub-arc second seeing of La Palma or Mauna Kea (although the jet stream affects that too) but we have to aim for making the most of the best nights our local climate provides, whether it be 1, 2 or 3 arc second resolution.

It is fair to say that it's best to observe every night it is clear, and especially after midnight if you possibly can. Going to bed early and getting up for a pre-sunrise observing session is really rather nice, as you get to see the first hints of the forthcoming sunrise and enjoy a peaceful time few appreciate. It does take a bit of getting used to, but ensures you can get to see some beautiful sunrises.

5. **Careful collimation:** When you see figures quoted for the resolution or magnitude limit for your telescope there is an unspoken subtext which says "If it is perfectly adjusted". The calculations used to determine the resolution and brightness of the faintest stars discernible using a telescope are generated by working on the firm assumption that all the optical components are perfectly aligned – otherwise known as collimated. As might be expected, this is often

not the case, particularly for instruments that are carried in and out of sheds and not permanently mounted. The act of gently placing a telescope in a car and driving a few miles can be sufficient to degrade the alignment to such an extent that images of stars are not small symmetrical dots and planets appear decidedly blurred.

So, before focusing, check your collimation by either using a Cheshire eyepiece, or a laser collimator, and then finish it off by tweaking it by hand while watching a real-time image of a star, to see which adjustment generates the tightest images. As with focusing, it is worth retesting occasionally during the night, particularly if you change from observing an object of high elevation to one that is low, or from an item in the south to one in the north. All these changes can cause subtle realignments requiring attention.

6. **Clean optics:** It's pretty obvious really that the best images will only be attained if the mirrors and lenses of your telescope are kept free of dirt and mirrors realuminized when they start degrading. Look for a faintly milky front surface, or look through the back of the mirror at a bright light and see if you can see more than a few pinpricks of light. This is not the place to go into a discussion of how to clean a mirror or lens – as can be required if some condensation drips onto the optics or a family of earwigs decides your secondary mirror is the ideal home (it happened to me). The rule with telescope optics is "Don't clean it" unless you really know what you are doing. Seek guidance before doing anything.
7. **Light-tight optics:** Just as important is checking that your telescope tube is light tight. It is essential that no extraneous light reaches the mirrors, eyepiece or cameras as it will appear as an artifact in the final image. Try taking an image with the light cap on and shining a torch at the focusing mount and mirror cell. If there are any significant leaks that will show them up.

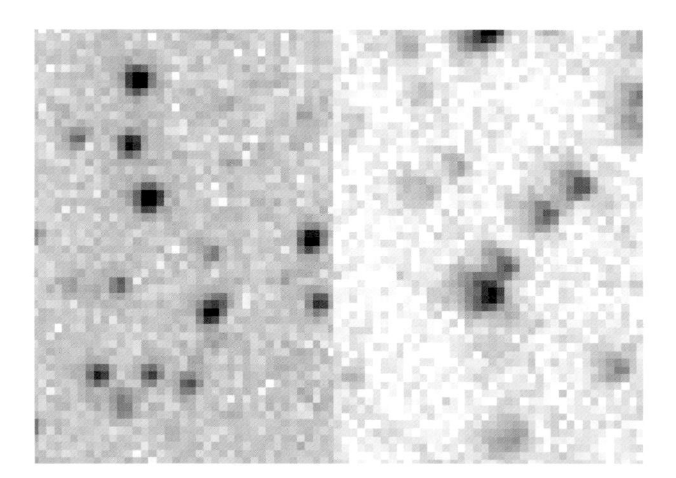

Figure 4.5. Examples of the distorted star images caused by careless collimation of the optics. Good collimation is on the left. Image credit: Grant Privett.

8. **Take lots of images:** There's a tendency among new imagers, especially those equipped with GOTO telescopes to find an object, take two or three images and then quickly move on to the next. It's undeniable that you will build up a portfolio of images very quickly and that will impress non-astronomer friends, but you will rarely create a great picture that way. This advice is doubly relevant to planetary imaging where you should bargain on filling tens of gigabytes of disk space a night. Personally, I would recommend never taking less than 30 minutes worth of images of a deep sky object and preferably 100 or 200 individual images if the exposures are shorter than 1 minute. Under bad weather conditions with only clear periods this could take several nights, but the results will be worth it. With a webcam it is best to ensure you always take thousands of images. It takes a long time to set up, so why not get the best you can from the effort?
9. **Ensure your telescope has cooled down:** One of the image degrading problems encountered when using a telescope is that when you take it outdoors it will almost certainly be warmer than its surroundings. This causes the air around the optics to move chaotically and can reduce the image quality to mush. Some people insist that Maksutovs and SCTs are immune to this but that's complete garbage. Even refractors are affected to some extent. The problem is most pronounced in larger instruments where the mirror is a substantial mass of glass and may take well over an hour to cool to approach the ambient air temperature. One way of encouraging the telescope to cool down so you can start to observe is to put it out as soon after sunset as possible, that way the period of poor imaging occurs while the sky is still bright. Alternatively, use a fan to force air past the mirror. The small 12 V fans commonly used to cool home computers can be very good for this purpose. If buying new, it's worth looking for a low-noise fan as they will generate less vibration. Mount the mirror on rubber band supports strung below the mirror cell to help absorb any vibration arising from the fan bearings. In recent years it has become increasingly common for SCTs and Newtonians to be fitted with such fans as optional extras.
10. **Avoid condensation:** At some times of year – it obviously depends upon where you live – you may find that water starts to condense on both the tube and optics of your telescope. When that happens you will find your images quickly become dim and the stars bloated. Initially, when the layer is thin, it may take some time for the degradation caused by the water to become apparent when looking at the screen – especially if the image scaling is automatic – so be sure to examine the optical surfaces every half hour or so. A hair dryer can be a useful accessory, but dew shields extending the telescope tube length or small heaters are very effective long-term remedies.

CHAPTER FIVE

What Is Best for ...

The sections below describe broadly what the best combinations are for observing a particular type of astronomical object. The options it selects should be viewed not as restrictive, but as an indication of a good starting point.

The Moon and Planets

Because of the ease with which they can be located and their great brightness, the quest to create a good picture of the Moon or a planet is frequently the first type of serious astronomical imaging anyone attempts. One of the great benefits of the brightness involved is that comparatively short exposures – usually a small fraction of a second – can be employed. This freezes the seeing, thereby permitting moments when the view is undistorted by atmospheric turbulence to be captured and preserved – along with a lot of images showing moments where the atmosphere caused blurring. Obviously, it's impossible to predict exactly when during a night the good moments will occur – so taking a long sequence of images is the best way to approach the problem. A secondary benefit of the great brightness of the planets and Moon is that inaccuracies in a telescope drive which would ruin a deep sky image (by turning star images into elongated unsightly blurs) can be ignored.

These considerations, combined with the advent of affordable webcam imaging has led to the situation where the standard of images created by astronomers from their back gardens is now staggeringly high, and popularity for this sort of imaging has mushroomed. It is likely, for example, that the number of fine images captured during the highly favorable 2003 or 2005 oppositions of Mars exceeded the total number of high-quality Mars images ever taken from the Earth. If you need convincing of how far planetary imaging has come in just a few years, try to find a copy of Patrick Moore's *Guide to the Planets* from the mid-1970s and examine the photographs it contains. The only examples that are passable by today's standards were taken with professional telescopes such as those at the Palomar Observatory. That's not to say people did not work hard to get those images, it's just that technology has moved on a long way, and the gold standard with it. Nowadays anyone who can afford to buy a driven 200 mm (8 inch) telescope and a webcam can take fantastic planetary and lunar pictures. Every year the standard seems to improve.

It is undeniable that some of the best planetary images of the last few years have been taken by the likes of Damien Peach, Antonio Cidadao, Eric Ng and Don Parker. As Damien Peach has acquired some exceptionally fine images while imaging from the UK, it cannot even be said that observing from island mountains is essential to obtaining really good images. What is essential is dedication and attention to all the details mentioned in Chap. 4: "Acquiring Images". It is also noticeable that many observers, who at one time employed individual R, G, B filters with a CCD, now employ a webcam such as the Philips ToUCam 840k Pro or the considerably more expensive Lumenera LU075. Other recent market entries, including the webcam-like cameras from Celestron and Meade (NexImage and DSI/LPI, respectively), also seem poised to make an impact.

The webcam allows the acquisition of thousands of images in only a few minutes and, when used in association with processing software such as *RegiStax,* can create a single image showing what the planet might have looked like if the view had been not been distorted by the atmosphere.

Figure 5.1. A very fine image of Saturn. This image is fairly representative of the current state of the art of ground-based planetary imaging and of this imager's superb portfolio of webcam-based work. The image shows the planet on January 23rd, 2006.
Image credit: Damian Peach.

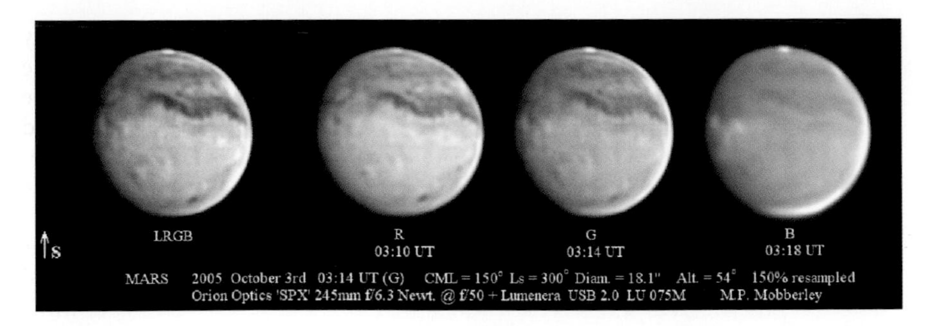

Figure 5.2. An image of Mars created by Martin Mobberley using a 245 mm (9.5 inch) Newtonian reflector operating at f/50. The image on the left is an LRGB composite created using color information from the other three images. Note how the images were acquired at roughly the same time to avoid smearing caused by the planet rotating.
Image credit: Martin Mobberley.

The planets should be observed with an imaging scale, which means one arc second is spread out over several pixels. Try to use a telescope operating with a Barlow lens at a focal ratio greater than 25. The type of telescope you use is less important than attaining a good focal ratio.

One factor we are overlooking here is that the planets rotate, bringing new territory into sight and taking other features away from view. This means that if your image sequence is too long, some features will be seen to have moved slightly and so will appear blurred on the composite image created. As all the planets rotate at different rates the time limit varies considerably. Similarly, the distance of the planets at opposition, and hence their apparent size, has a profound impact, but reasonable limits are of the order of 7, 1.25, 3.25 minutes for Mars, Jupiter and Saturn, respectively, under ideal seeing conditions. On a night of poor seeing these may be relaxed somewhat to, perhaps, 10, 2 and 5 minutes, but the result will start to degrade quickly (as if poor seeing wasn't bad enough) as the length of sequence used grows – particularly when imaging Mars and Jupiter.

Obviously nothing – apart from cloud, disk space and fatigue – stops the observer creating a number of such sequences during the course of a night. The other planets have rather fewer distinct details visible in amateur telescopes, so detecting anything on Venus, Mercury, Uranus and Neptune is widely counted an achievement. It is worth noting that the potential of a good telescope combined with a webcam is such that some amateurs now pursue detail on tiny targets such as Jupiter's moons.

Deep Sky

Unlike planetary imaging where photons are available in abundance, most deep sky imaging is about collecting as many photons as possible. Inevitably, many observers have tended toward long exposures and the use of auto-guiders whenever available – and the stacking of shorter exposures when not.

Traditionally, color deep sky imaging has tended to mean that a monochrome CCD must be used in conjunction with a filter wheel, but one-shot color cameras such as DSLRs and some CCDs are popular for their ease of use, increasing affordability and convenience. Webcams, by comparison, often provide too small a field of view and only permit the use of short exposure lengths. For deep sky work, one-shot color cameras can be viewed as less sensitive than the monochrome alternatives, but have the great virtue of providing images where there is no need to align and combine the red, green and blue images.

As to telescopes, a modest instrument, even a Newtonian 114 mm (4.5 inch) reflector can create impressive images if it is on a good mount. A popular instrument is a Newtonian with a focal ratio of around 6 – especially if a parabolic corrector lens such as the Televue Paracorr is used to ensure that a wider field of view still has pin-sharp star images. Newtonians are often favored above SCT because, inch for inch of aperture, the Newtonian is considerably cheaper and effective despite the diffraction patterns caused by the "spiders" that physically support the secondary mirrors.

That said, an SCT will perform pretty much as well as a Newtonian and frequently benefits from a GOTO mount. Certainly both are cheaper than the increasingly popular apochromatic refractors which provide a good sized field of view, crisp star images and portability – but at a substantial financial cost.

Aurora and Meteors

For aurora and meteors the best imaging option is a wide-field lens attached to a DSLR. Color is an essential element of any aurora picture as an otherwise green display can suddenly flare up with rays or curtains of blue or red light. It also serves to

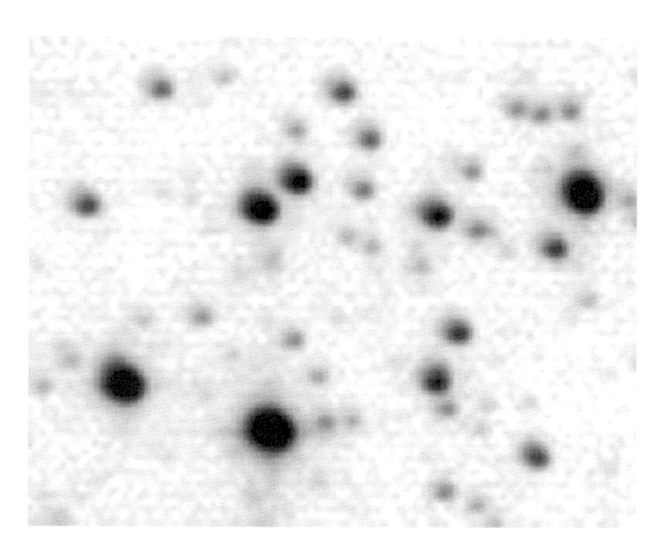

Figure 5.3. Star images away from the central "sweet spot" of the focal plane – out into the region where coma is present. For some instruments, corrector lenses can provide a solution by increasing the field of view over which star images remain circular. The effect can also appear if the telescope is poorly collimated. Image credit: Grant Privett.

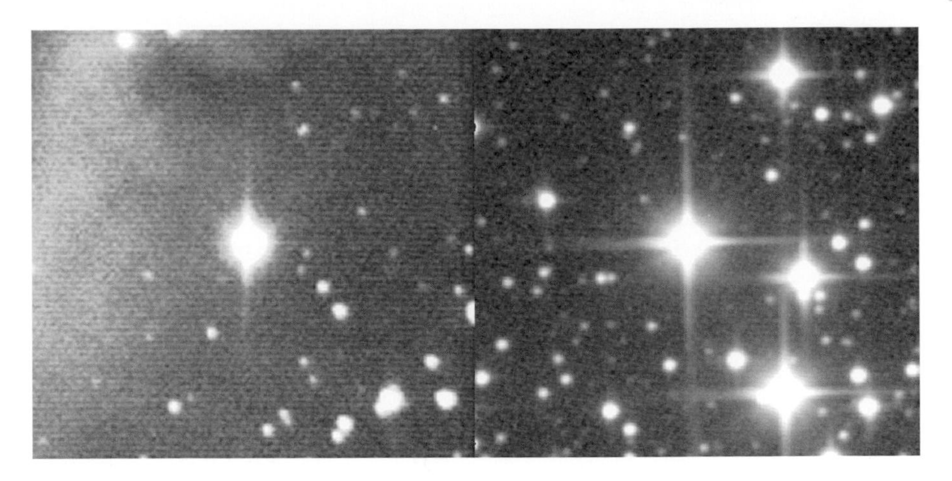

Figure 5.4. These examples show the effect on an image of having one (left) or two (right) supports for your secondary mirror in Newtonian reflector. Image credit: Grant Privett.

make meteor shots much more attractive by allowing the color of the stars to be seen. Something like a 28–50 mm (1.1–2 inch) lens will show a large chunk of sky, which is essential as aurora and meteors can extend from horizon to horizon. For aurora, by taking a sequence of images at 5 second intervals you can make a movie showing how the curtains or rays changed as time progressed. You will probably find that exposures of the order of 1–10 seconds in duration are enough depending upon how strong the display is. Remember a strong display can make the whole sky brighter than a 2nd magnitude star, so it's very easy to saturate a sensor.

Expect to spend a considerable amount of time tweaking the color balance and brightness levels of the images to ensure that the aurora looks the same as it did in the sky, and to bring out faint detail.

For meteors, you can take multiple images of the same chunk of sky and – if no meteor is seen – later stack them to create an attractive constellation picture. Inevitably, you will probably find that you have also recorded several satellites, so it's worth recording the specific date and time of each image to make identification possible. If you observe a meteor pass through the field of view the imaging can be stopped immediately before the background becomes bright. Also, if a meteor trail forms, a further image can be taken to show how the trail was distorted by the tenuous upper atmosphere winds. Commercial satellites such as *Iridiums* or the *International Space Station* are a good source of practice – but remember meteors are much faster. For scientific observation of meteors it's worth investigating low-light video surveillance technology and the Mintron camera.

The Sun

The great thing about the Sun is that there is certainly no shortage of photons available and a humble 60 mm (2.5 inch) refractor or small Maksutov such as the Meade ETX can produce images with quite a good resolution, even allowing surface granulation to be seen if the air is steady. The obvious camera types to employ are DSLRs and webcams. Either of these can be used if the telescope is fitted with an inexpensive Mylar solar filter that cuts out all but a millionth of the sunlight and will give a good representation of the Sun in white light, i.e. as seen by simple eyepiece projection. Be sure to buy only from reputable suppliers such as Baader and Thousand Oaks Optical – holographic Christmas wrapping paper will not do. The filter will allow detail within sunspot groups to be readily observed, and is especially successful when imaging eclipses or transits of Venus or Mercury.

However, it is becoming increasingly popular to observe the Sun in H-alpha light using one of the new breed of specialist solar telescopes, such as the highly respected Coronado that employ an etalon – a very narrow-band filter that only passes light of a very precise color. These show incredible additional detail such as solar prominences, faculae and filaments. To get the best out of such an image you need a camera with a large sensor. This allows the whole Sun to be captured in one go, but often much of the Sun is fairly bland and a smaller sensor can be just as successful by framing a small region of interest containing a group of spots or small prominences. A manual focus is another essential when imaging the Sun, as some auto-focus systems may struggle when it is targeted.

An unsharp mask and other sharpening routines will be of particular use with the Sun. You may also find some potential in applying deconvolution techniques to extract all the fine details.

Comets

Comets present several problems for the imager. The most readily apparent is that spectacular comets can be quite large, spanning as much as 60 degrees across the sky (Hyakutake from 1995 was a fine example) but in addition to that they can move quite appreciably during the course of an exposure. So if you are using separate color filters to image them you will find that when the images of the comet are all aligned and stacked it will lead to an image containing some very strangely colored star trails. For this reason DSLRs and one-shot color cameras are a popular tool. Great care must be taken to ensure that the core of the comet does not saturate the sensor as this will spoil an image – as an example, the core of Comet Hyakutake saturated a TC255 sensor on a 200 mm (8 inch) telescope in 2–3 seconds.

However, for slow-moving comets or those with faint cores – which are pretty much always small – a simple unfiltered monochrome CCD does rather better, as you then need to collect all the photons available and not sacrifice any of them to filtering.

One thing worth looking for in comets is detail in the tail, so expect to spend some time with the histogram (of which later) adjustment options of your software to bring out detail in the tail without a brilliant core totally dominating the image.

More esoterically, you will often find the Larson–Sekanina and other types of radial gradient operators implemented in your image processing package. These help show where faint filaments of tail or coma may be present and highlight the detail. Unfortunately, the output images they create can look seriously weird,

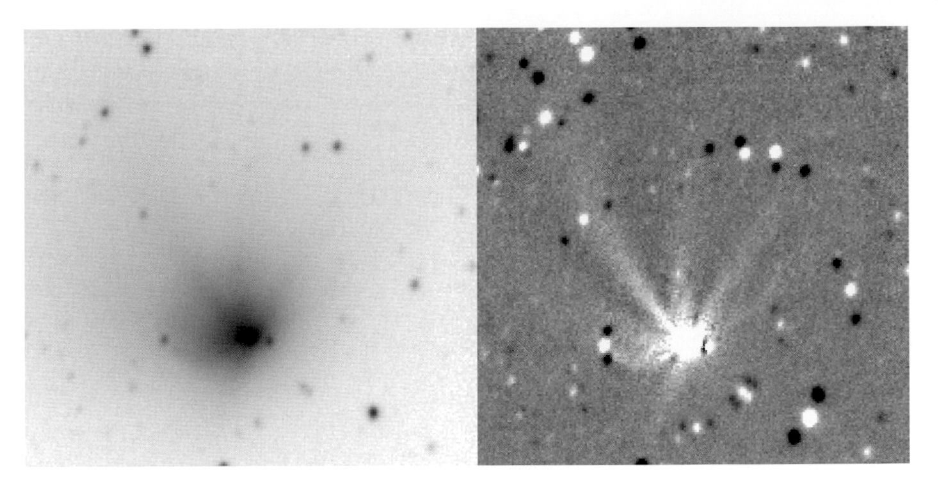

Figure 5.5. The processing of imagery of comet Hale–Bopp on September 15th, 1996 showing an unprocessed negative view, and after the Larson–Sekanina algorithm was applied. Note the multiple jets from the core. Image credit: Grant Privett.

especially if false color is applied. The images created can be a little difficult to interpret, but are often the only way of detecting such detail and so are important in the serious study of comets.

Another feature that can sometimes be teased out of images of the inner coma of brighter comets is shells of gas thrown off by the core. An unsharp mask is a popular approach. Particularly fine examples were seen in comets Hyakutake and Hale–Bopp. Surprisingly, careful processing may be necessary to show the numerous shells that were so obvious visually in the eyepiece. Imaging these phenomena really does require that rigorous image reduction techniques are closely adhered to.

In images of comets it is also very worthwhile, making sure you carefully remove the sky gradients. Faint filaments within the tail can easily be hidden or masked if the gradient is not fully corrected.

CHAPTER SIX

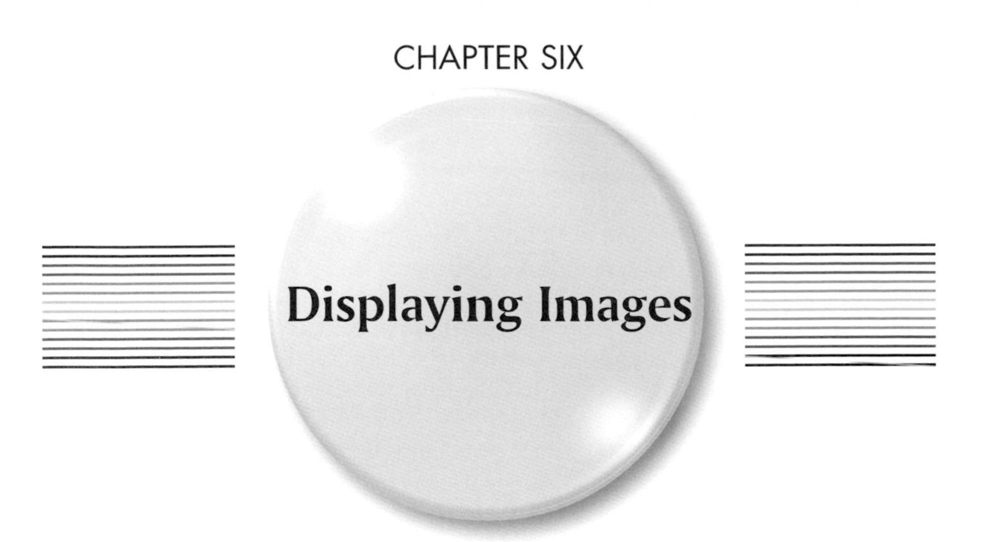

Displaying Images

One of the big surprises of looking at your images on a computer is that the view you get will vary considerably, depending upon a number of factors. You might think that all displays are much the same, and for purposes like word processing or email that's pretty much the case, but in the more demanding arena of image processing the differences become more important. What you would ideally need is a display system which could accurately display at least as many gray levels or color shades as your image contains. A 16 bit monochrome image could contain 65,536 shades of gray, and a 48 bit color image could contain billions of individual hues.

Fortunately, all is not lost, for whilst we might intuitively prefer our displays to present everything our data has to offer, it would be a bit of a waste of effort as our eyes are not capable of discerning all the subtleties. In general the eye can discern only a few dozen shades of gray. In comparison, the eye can detect a far greater number of colors so we need to ensure the display can reflect that. Modern LCD/TFT monitors are very good, use less power, take up less space and weigh less than a CRT monitor. Yet even now, as CRTs are starting to disappear from catalogues in the same way that VCRs did, for the most demanding purposes a large, high-resolution, good-quality CRT is still a great buy.

To ensure that the display is configured properly, many astronomers use a display test image showing the complete range of gray levels from black to white. When this is displayed they adjust the image brightness and contrast to ensure both can be clearly seen. Other test images can be used to configure the color balance and make sure white objects really do look white. This latter process used to be essential, but many modern monitors appear to be very stable from month to month and require little alteration after being first set up.

Another major consideration is the aspect ratio of the display. With CRT technology it was possible to vary the image display scaling so that a square image always looked square on the screen. This is not the case for TFT display users who will have noticed that while their screens will usually operate in a large number of modes from 800 × 640 pixels upward, only a few of the modes will display a square image as square. This effect is most apparent when looking at images of people – they will have impossibly long or squat bodies. Consequently, after some experimentation with your *Windows* Control > Settings > Display options, you may find that the best screen mode for displaying images is not the setting you would otherwise have employed when using your computer. If that's the case, you might want to set up a user account that has those settings, and use it only for astronomical image processing to avoid having to switch back and forward between screen display settings.

The size of a monitor is an important consideration. As someone brought up on the 230 mm (9 inch) screens that came with the Commodore PET computers of 1980, I can vouch for the fact a 482 mm (14 inch) screen is highly usable, even at high resolution. But I would suggest that whenever possible a larger screen should be used, especially when multimegabyte images are being examined. Currently 432 mm (17 inch) LCD screens are becoming affordable, and it can be imagined that this will become 533 mm (21 inch) within a couple of years as the monitor market will be influenced by the drive toward bigger and bigger television screens. LCD screens also have the virtue of costing considerably less to run over their lifetime than a CRT, which can be worth considering if you leave your computer turned on at all times – standby mode often saving less power than you might expect.

The other important way of displaying images is to print them out. The ability to do this has improved greatly over the last few years, and now even comparatively inexpensive printers can do a very fine job of representing our images. Of course, printing astronomical images can be rather heavy on the consumables – it certainly eats the black ink – and when making a print you intend to keep, it is best to take care to use good-quality photo paper. Epson, Canon, HP, *et al.* now all sell such paper at a reasonably sensible price – it's funny how the word takes on a different meaning when IT equipment is discussed – and printers specifically aimed at the photographic market. It is also best to clean the printer heads first and to use the highest quality settings – usually also the slowest – on the printer. Even then do not be surprised if the quality is not quite what you would expect from a photograph printed professionally. It may be worth putting your better images onto CD and having them printed out professionally, or even by high-street shops.

A further concern is for the durability of prints. I have a black and white print of my great grandfather which is now over 100 years old and looks fine albeit a little scratched. Yet, several computer printouts I own of astronomical images taken in the 1990s have now seriously faded as a result of sunlight or being kept in slightly damp conditions. Fortunately, I have retained the original digital images so regenerating the images is not an issue, and it should become still less so as time goes by and ink technology develops further. Despite this, for really top quality results it is still worth using a professional Cibachrome printer to obtain sharp, strongly colored and shiny results that will take some years to dim. Unfortunately, they are still a little expensive to use, but, for a picture you intend to hang on the wall or give away, they are pretty much unbeatable if produced by a reputable company.

CHAPTER SEVEN

Image Reduction

As mentioned previously, the term "image reduction" covers the processing that must be applied to a set of images to remove all the artifacts generated by the detector and telescope. These processes must be undertaken for every object you wish to image. Under some circumstances shortcuts may be taken, but it's easiest to just collect lots of images and do the job properly every time. They define how successful the subsequent image enhancement stage will be. Do them properly and reap the full rewards later.

The Importance of Dark Frames

Probably the most important single stage of image reduction is the process of dark subtraction. It is quite possible to ignore dark frames and nonetheless obtain good results, but you will have to work a lot harder to do so, and you will always know the image you ended up with could have been better! It is best to simply accept that taking dark frames is an essential part of image processing and get used to employing them whenever you image.

The inevitable question, then, is what is a dark frame? Simply stated, a dark frame is the image your camera would have provided if the telescope had been capped and all light prevented from reaching the light-sensing chip. Now, if you were sure your telescope was light tight you might expect that the camera would record nothing, and in an ideal world that would be the case. But, even today's cameras suffer from a number of imperfections, and the images they generate will contain artifacts, so that even in complete darkness the image you get back from your camera will show something similar to the examples in Fig. 7.1.

In general, a dark frame will look very similar to the picture you obtain from a television that has not been tuned in to a station – largely random noise.

Figure 7.1. A typical dark frame for a sensor made in 1994. The image has been scaled to allow the noise to be seen. On the original the two bad columns are very faintly visible as anomalous compared to their neighbors. The remaining noise is quite random. Image credit: Grant Privett.

But, depending upon the exposure length and the camera type employed, it may display some columns that appear to stand out or are dark, a gradient between the top and bottom of the image or even a brightly illuminated image corner. The columns and brightening at an image corner arise from a combination of imperfections in the manufacture of the sensing chip, and also from the manner in which the charge in each pixel of the chip is read out – they are a function of the chip. Do not worry, we can deal with them.

CCD camera manufacturers often minimize the amplitude of the random noise by reducing the temperature of the CCD chip – the majority of the signal being thermally induced. Cooling the camera chip works extremely well, as the relationship between CCD temperature and dark noise causes the noise to fade by a factor of around 2, for every drop in temperature of 5 °C (9 °F).

So, if you generate a dark frame at an ambient temperature of 15 °C (60 °F) – summer – and subsequently compare it with another taken at a temperature of 0 °C (32 °F) – winter – you would find the noise in the latter is but an eighth as bright. Because of this dramatic reduction, many CCD manufacturers cool their products via single-stage thermoelectric Peltier coolers (similar to those used in portable 12 V picnic or beer fridges) and operate the CCD at a temperature of 30–35 degrees below the ambient temperature. Single-stage Peltier cooling is so successful that a camera might be able to operate at a chip temperature of –35 °C (–31 °F) in winter and in summer attain a temperature of –20 °C (–4 °F) with the difference arising from the difference in seasonal ambient temperatures. Today, the excess heat from the camera head extracted by the Peltier is often dissipated by the use of cooling fans, but in the 1990s water/antifreeze cooling systems were sometimes used and buckets full of antifreeze were a necessary accessory for the serious observer.

You might suppose that if some cooling is good then lots must be better, and manufacturers would therefore provide all cameras with several coolers (and

Figure 7.2. A typical dark frame for a CCD sensor made in 2003. Note the brighter upper left corner where a signal caused by the on-chip amplifier can be seen. This will be easily removed from images by dark subtraction, due to its small amplitude. In the remainder of the image there are a small number of hot pixels. Overall, the noise from the on-chip electronics and thermal sources is considerably reduced compared to its 1994 predecessor. Image credit: Grant Privett.

perhaps some antifreeze in a bucket) working together to provide more cooling. Well, some companies have indeed adopted that strategy, but multistage cooling not only requires more power, but has been known to cause problems. The intensely cooled cavity containing the sensor chip can become so cold that condensation forms within the camera or upon the surface of the camera window. Consequently, the designs employed for commercial cameras have to balance several factors, not least the fact that doubling the current supplied to the coolers does not make the chip twice as cold.

Some cameras allow you to select the temperature at which a camera operates, while others operate their coolers working on a "flat out" uncontrolled basis. It is more important for the image reduction that the temperature of the chip remains stable, than the temperature maybe being a degree or two lower yet varying. Those observers using single-stage unregulated cooling should consider using a portable digital thermometer while observing, to monitor the air temperature for sudden changes – as these can very quickly occur when cloud passes through – and be sure to take further dark frames whenever significant temperature changes have occurred.

Unfortunately, a lot of this discussion of cooled sensors does not apply to most commercial webcams, and, at the time of writing, I am unaware of any widely used DSLRs that have cooling options. It goes without saying that users of webcams and DSLRs must be especially vigilant to ensure that dark frames are always taken when long exposures are being acquired.

Notable exceptions to the rule regarding the cooling of webcam-like cameras are the SAC and Orion Starshoot ranges of cameras, that have been modified specifically for the astronomy market to allow some cooling of the sensor, and

Figure 7.3. DSLRs contain sensors that are uncooled. Often, in long exposures, the dark images will show an artifact feature generated by the readout electronics. Such features change in appearance depending upon the ASA setting used, but will normally vanish after subtraction of a suitable dark frame. The contrast has been adjusted to make the feature more visible. Image credit: Mark Wiggin.

exposures of lengths normally associated with CCDs. It is possible to undertake some of the hardware modifications yourself – advice can be found on the web – but anyone contemplating such action would be best advised to consult the QCUIAG website in the first instance. You can also wave your camera guarantee goodbye.

The bottom line is to always take dark frames.

Taking Your Dark Frames

No matter which type of camera you use, it is essential that you turn on a cooled sensor early in your set up so that it can reach a stable temperature. Even an uncooled camera should be taken outdoors early on to allow its temperature to stabilize with the air and telescope. The time taken can vary from 5 minutes to nearly 30 minutes depending on the camera. Obviously, care must be taken to ensure it does not get dewed up – losing a night that way is really frustrating. Keep a hair drier to hand.

After first taking a few trial images to determine the exposure length you will need to use for the current target, make the telescope light-tight and take a series of 5–10 dark frames of the chosen exposure length. If your camera temperature is regulated and steady you need only do this once, either immediately before or

Figure 7.4. Before and after dark subtraction. The image of comet Tempel1 on the left has not been dark subtracted. The image on the right has. Image credit: Grant Privett.

after the images are taken. But if you are using a camera with an unregulated cooler it is best to take a set of darks at the beginning and at the end of imaging the object, and combine all of the dark frames to create a master dark frame.

A strategy employed by those obtaining multiple images using cameras with no cooling is to take groups of dark frames at regular intervals throughout the night. For example, if you are taking 100 × 30s images of a faint cluster, you might take 10 dark frames at the start, 10 more after the 33rd image, a further 10 after the 66th and then another batch of 10 darks at the end. In this way images 1–33 could be reduced using the first two sets of dark frames, images 34–66 using the second and third sets of dark frames and images 67–100 using the last two sets of dark frames.

You will have noticed that in taking a total of 40 × 30s exposures we have set aside at least 20 minutes of clear sky for imaging nothing! This may seem a terrible waste of clear sky, but it would have been a far greater waste to have gone to the effort of taking 100 × 30s images and still create a poor image by neglecting the need to get a good estimate of the dark frame. It can take a considerable effort of will on a pristine night to take the final set of dark frames before moving on to another object. Dark frames are the "ha'p'orth o' tar" on the route to good images.

Creating Master Darks

Up until now I have blithely stated that we will process the images we obtain using our dark frames. Before we can do this we must create a master dark frame from our sets of darks. We cannot simply pick one of the individual dark frames at random and expect to create consistent results. If you examine two dark frames closely – even two taken consecutively – you will find that they differ, if only slightly. The master frame you create is an estimate of the result you would receive if you could take many hundreds of dark frames and identify from them the most typical representative result.

A master frame is created by first selecting all the darks you are interested in, choosing a method to be employed by the software and then letting the software run. The processing required to create master dark frames is extremely simple, and will be pretty much limited by the time it takes your computer to read all the dark frames from disk. Realistically, a few seconds to a minute or two will suffice for all but the biggest cameras. You will probably be offered a number of options regarding the manner or algorithm that is to be used when the dark frames are to be created. One will almost certainly be "Mean" or "Average". This option will involve examining the same pixel on all the images, calculating the sum of the values found, and dividing by the number of images to provide a mean or average value for a given pixel. The process is repeated for all the image pixels in turn. As a method it has serious faults and, unless you have very few dark frames (less than five), the "Mean" option is probably best avoided. The other choices you are likely to encounter will be "Median" or a "Clipped Mean". These options employ slightly different approaches to minimizing the difference between the master dark frame you are creating and an ideal master dark. The "Clipped Mean" looks at the same pixel on all the images but ignores the highest and lowest value(s) it finds when calculating the mean value returned. Consequently, it is less sensitive to noise within the pixel count value. The "Median" approach looks at the same pixel in all the images, puts them in order of increasing value and then selects the middle one – the median value. Under ideal conditions, the resultant value will be even less sensitive to the influence of noise – in real life the median and clipped mean approaches may give very similar results. It may be worth creating several master dark frames and comparing the results to see which works best for you. The method described will also work well for bias frames.

You might imagine that with a really good camera you could use "Mean" because the pixel count noise would be so small as to be no longer an issue, but even the finest cameras suffer from the effects of "cosmic ray" hits. Showers of atomic particles are generated high in the atmosphere by incoming cosmic rays of enormous energy. These particles cascade down through the atmosphere to the ground – and your camera. Most of the resultant cascading particles do not reach the ground – more cosmic rays are detected at the tops of mountains than at sea level – but some do, and they generally interact with a sensor chip to produce intensely bright elongated spots or short tracks, spanning a few pixels.

If a bright cosmic ray hit is present in a dark frame stack that is processed using a "Mean" algorithm, the resultant master frame will be seriously flawed. The "Median" and "Clipped Mean" algorithms will normally ignore the effect of the cosmic ray and produce a better final result. It is salutary to examine images taken by the *Hubble Space Telescope* (HST) and consider the enormous number of cosmic rays that have been processed out of each attractive HST picture released by NASA.

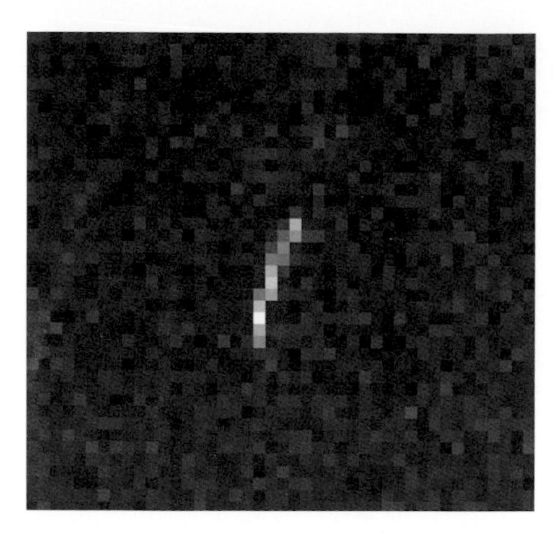

Figure 7.5. An example of a cosmic ray event. The majority are shorter than this at sea-level. Image credit: Grant Privett.

Figure 7.6. An example of a *Hubble Space Telescope* image before and after image reduction and cosmic ray suppression. Note the large number of cosmic ray hits. Image credit: NASA/ESA/Hubble Heritage Team.

Flat-Fields

We have seen that the collection of dark frames is essential to facilitate the removal of sensor-derived image artifacts from our images. Unfortunately, there are still more sensor artifacts that need to be addressed and the majority of these are handled via a flat-field frame.

The first of these artifacts is caused by dust within the telescope, not dust particles that are moving in the air currents, but those that have – inevitably – settled onto the surfaces of mirrors and lenses, or the window in front of the sensor. These particles, in essence, cast doughnut shaped shadows onto the sensor, and prevent some of the light from reaching the pixel it would otherwise have contributed to.

The size of the dust doughnut varies considerably, depending upon how large the dust particles are and how near to the focal plane they lie. Simply put, a dust particle on the window directly in front of the sensor will cast a shadow a lot smaller – but darker – than one that arises from dust on a Barlow lens, filter or secondary mirror. From this, it is clear that the place to start in reducing the need for flat-field images is by ensuring the optics of your telescope and camera are impeccably clean – something most of us ensure near-religiously anyway.

The second artifact that flat-fielding helps to counter – we must remember that nothing we do will perfectly restore the image – arises from an unavoidable imperfection within the sensor. Despite the best endeavors of chip manufacturers, it remains impossible to create a large grid of light-sensitive pixels and ensure that they are all individually of the exactly the same sensitivity or quantum efficiency (QE). In practice, each pixel of a sensor is usually very similar in sensitivity to all of its neighbors.

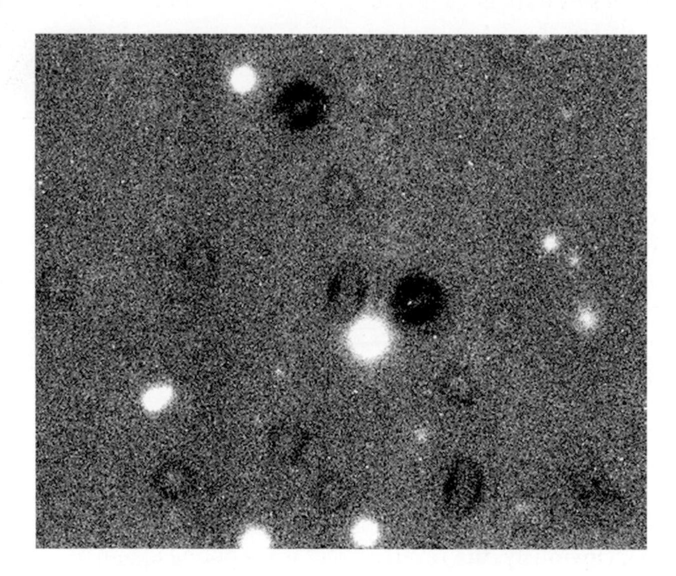

Figure 7.7. A typical flat-field image showing the "doughnut" pattern characteristic of dust particles somewhere within the light path. The size varies depending on the proximity of the dust to the focal plane. If a flat-field is not employed to correct this problem the resulting image will be entirely unusable. Image credit: Grant Privett.

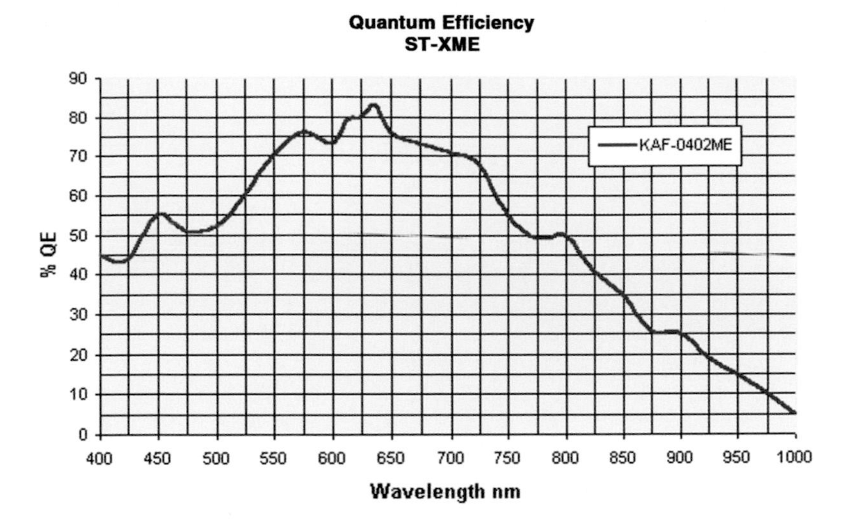

Figure 7.8. The sensitivity or quantum efficiency of a typical CCD as a function of wavelength or color. At 635 nm roughly 85% of photons contribute to the image created.
Image credit: Santa Barbara Instruments Group.

You will find that some pixels of the image are consistently much brighter than others in the flat-field frame – these are generally known as "hot pixels" and will record themselves as fully illuminated even in short exposures. Similarly, you will find some pixels are consistently less sensitive than the rest and are – no surprise here – termed cold pixels. Fortunately, chip fabrication processes have improved enormously over the years and the number of outstandingly hot and cold pixels has diminished sufficiently that it is now not uncommon to have less than 5–10 per million pixels. Older cameras may have 10 times this amount. There is no widely accepted hard and fast criterion for determining whether a pixel is hot, but a simple starting point is to process a single image fully and see how many pixels stand out obviously from a dark background.

Flat-fielding is accomplished by imaging a target that is entirely flat, i.e. possessing no brightness variations in its appearance. Alas, the universe does not provide many of those and so, sometimes, we have to improvise.

One strong contender is the clear twilight sky after sunset or before sunrise. The simplest way is to set a camera exposure of (say) 3 seconds and image the sky directly overhead continuously from the time the sky starts to darken. When you begin all pixels will be strongly saturated and uniformly white images will appear where noise can be seen. You can then start collecting sky flats – but be sure to stop before the signal drops below one-third of your saturation level (21,845 in a 16 bit camera). Obviously, the reverse process is applied at sunrise. The downside is that this requires that the sunrise or sunset is clear. Alas, for some of us the skies may not clear for more than a few hours at a time. For others, clear skies may not materialize until well after dark, and vanish into clouds or mist again well before the first vestige of dawn appears.

Figure 7.9. Another instance of the unsuccessful removal of a sky gradient. This time the removal of a model has revealed a fault in the flat-fielding – note the large dust doughnut at bottom left. Image credit: Grant Privett.

To get around this problem an alternative light source can be sought and used. It is possible to construct a light box that fits on the end of the telescope and provides an evenly illuminated, pristine white screen that the telescope images. The trick here is to ensure that the illumination really is highly uniform, so that variations in the flat-field frame do not reflect differences in illumination of the screen rather than the sensor. The details of construction of such a box are beyond the remit of this book but many examples can be found on the web.

One alternative source of flat-fields I have used has been passing clouds. If you live near a large town you will probably have noticed that at night clouds can be very brightly lit and will have a near uniform surface. I have, in the past, used a well-illuminated cloud-covered sky as a flat-field. These days I live in a darker location and employ a rendered concrete wall that is illuminated by diffuse light from an upstairs window. It's quite a good compromise – as long as no cars pass by and illuminate it irregularly through the hedge!

Okay, so having found some source of a flat-field image, what do we do with it? Well, the starting point is to note the exposure you employed for the flat-field and to collect a number of dark frames of that exposure length – 10 should be enough. Obviously, the downside is that we need to collect at least two sets of dark frames every night, one set for the dark frame we take and another for the flat-field images. Unfortunately, there is no way to avoid this as, whilst the pixel sensitivity will not vary significantly from night to night, the distribution of dust certainly can.

It is worth mentioning at this point that you should, if possible, devise some way – such as marks or paint blobs on the focusing mount and your camera – to ensure that the camera you use is always orientated the same way when on the telescope. This will help ensure that the dust doughnuts in the flat-field image from one night line up with images taken on other nights. This is a great convenience if it starts to rain or snow after you have taken images but before you have had a chance to acquire flats.

Processing Your Flat-Field

So, the night is clear, the sky is darkening and you have rushed outside and set up. Once the telescope is roughly set up, point it to the overhead position, insert your camera and select an exposure of (say) 5s. Then while the sky darkens start taking a series of images. At some point during that sequence the sky will darken sufficiently that the sensor is no longer saturated. As you watch the exposures come in you will see that the average value of a pixel within the image drops rapidly. Within a few minutes it will pass from fully saturated to a mean value below half the saturation level (65,535 in a 16 bit camera) and then ultimately to the point where the sky has darkened sufficiently for stars to appear in your images. Once that happens, stop taking images, seal the telescope up and take at least 10 dark images of the same exposure length.

When you have a chance to process your images, throw away those where the image background is below 25% of the saturation level or above 75% of the saturation level. Then take the dark frames you collected for the 5 s exposure and create a master dark as described previously.

Once that has been generated, subtract it from each of the flat-field frames. The images that now remain should then be combined into a master flat frame – as with dark frame creation you will probably have several options to choose from: "Mean", "Clipped Mean" and "Median". In the case of flat-fields taken at twilight with rapidly changing illumination levels use the "Mean", but for images of light boxes, clouds or walls use "Median" or "Clipped Mean" or similar.

The ideal master flat-field image will be the result of using flat-field images exposed to ensure a mean pixel count within the range of 25–60% of the pixel saturation value – broadly 16,000–39,000. That should be sufficient to flat-field imagery for all but the most demanding purposes and avoids the regions of the chip where its response to light is not linear.

Now you have generated the master flat-field image you can apply it to your image set.

The processing chain we have outlined so far goes rather like this:

0. Take your images, darks and bias frame.
1. Create master dark frame
2. Create master flat-field dark frame
3. Subtract the master flat-field dark frame created in stage 2 from all the flat-field images.

4. Create a master flat-field using the dark subtracted flat-field images created in stage 3.
5. Subtract the master dark frame from your images.
6. Flat-field your images using your master flat-field.

At the end of stage 4 we have reached the point where we have removed the vast majority of artifacts within the images. There remains much more we can do, but the most tedious part of image reduction is now behind us. It can all look a little convoluted, but do not be put off. It quickly becomes second nature, and after seeing the results you will be surprised that anyone ever attempts to skip these bits.

You may be wondering how the flat-fielding works. Mathematically speaking a division process is applied. A simpler way to look at it is to understand that whenever the image processing software looks at an image pixel it asks itself how bright the corresponding pixel was in the master flat-field images. If the pixel in the master flat-field was below average, the pixel in the new image is made a little brighter to compensate. If, on the other hand, the corresponding pixel in the master flat-field was brighter than the average pixel in the master flat-field, the pixel in the output image is made slightly dimmer to compensate. By applying this sort of operation to every image pixel, the dust doughnuts and variations in pixel sensitivity described earlier should be largely removed. The correction will not be perfect but will be quite good.

The idea and practice of subtracting lots of darks from other images can be extremely tedious – watching paint of any color drying is far too exciting an exercise by comparison. Fortunately, the individuals who write most of the popular astronomical image processing software, take pity on us, and provide batch processing approaches to this task. These make life much easier as all you need do is to choose the images that should be used to process the darks, bias frames, flat-fields and dark frames for flats. Usually a drag-and-drop interface is used to allow the files to be selected from your folder/directories and automatically processed according to the options you identify. Often the default options in these systems are quite sensible, but once you become more experienced, revisit the option menus and try variations. Programmers are far from perfect, and can occasionally choose an option that worked well for their suite of test images, but which messes things up for the general user.

When using a batch process it is advisable for any user to start small, adding to their batch process slowly and testing the output images at each stage. For example; create a master dark from a few dark frames and then look at it to check it is okay, only then consider indicating the images the master dark is to be subtracted from. Once you are content that this part is working properly, start working on creating your master flat. You can save a lot of effort working this way.

Bias Frames

I mentioned previously that the temperature of the sensor has a large impact upon the pixel count values found within the dark frame, but equally important is the duration of the exposure. The thermal contribution to the dark frame signal *is* directly proportional to the exposure length, while the part of the dark frame signal arising from electronics that reads the signal from the sensor chip *is not*. Perhaps the best way to think of it is that there is some signal you get just because you read out the sensor chip, and some because of the heat within the sensor created random pixel counts. It is possible to see these two components separately. Set up your camera in complete darkness and take two pictures. One with the shortest exposure your camera will allow (typically less than 0.01s) and the other of 60s or more duration. Now, the first exposure – better known as a bias frame – will show a fairly small signal and possibly some of the variations we mentioned earlier – such as a brighter corner with small-scale random noise superimposed and, possibly, even a top to bottom gradient. An example is shown in Fig. 7.10. If you instead look at the 60s exposure you will find that the pixel counts are considerably greater, and the bias frame signal is pretty much overwhelmed. To create an image containing only the thermal contribution to the image you will need to subtract the bias frame from the 60s exposure. You then have an image containing 60s worth of thermal signature without the influence of the read-out electronics.

Such an image may, at first, seem pointless or, at best, of only academic interest, but it isn't. If you take some long (say 120s) dark frames and bias frames at the start of every night they will act as insurance if you forget to take any darks of any other exposure later on. For example, to create a 20s dark, average the bias frames and subtract the result from the average of the 120s dark frames. Divide the resulting image by 6 (which is the ratio of the exposure time – i.e. 120s/20s) and then add the average bias frame again. This gives you an image that is equivalent to a 20s dark frame. Similarly, a 30s dark would be created in the same way but using a division by 4 instead. This can be a life-saving technique on nights when you image lots of different types of object and use very different exposure lengths. In the heady excitement of locating targets and acquiring images it's very easy to forget a few darks – I certainly have, I only seem to realize when I am indoors again turning the kettle on – but the simplest thing to do is *always* take darks.

Figure 7.10. A typical bias frame for a sensor made in 1994. Note the bright edge and the two bad columns. Image credit: Grant Privett.

Figure 7.11. A typical bias frame for a sensor made in 2003. Note the random appearance of the noise, and the absence of bad columns seen on earlier generation chips. The amplitude of the deviations of pixels from the mean value is now five times less than it was in 1994. Image credit: Grant Privett.

Interestingly, some observers with temperature stable camera sensors always take a number of bias frames and a number of long dark frames and use those to create any darks they need and never take normal darks. The statistics of the process mean that it can be an advantageous approach. But, for the beginner, it is worth becoming comfortable with the rest of your image processing and enhancement, before revisiting this approach to see if it shows real benefits with your own images.

Image Transformations

An image transformation (sometimes known as registration) is a process applied to an image that makes all the stars in one image line up with those in another image of the same chunk of sky. Transformations may take several forms. If different cameras have been used to take images there may be a scaling transformation required as the pixels in, for example, a Starlight Xpress MX716 (8.2 × 8.4 microns) are not the same size or shape as those within an SBIG ST7-MXE (9 microns square).

Alternatively, if the same camera has been employed to image the same object on different nights, then the pixel scale will be the same but the images may be rotated with respect to each other. So, image rotation will be required.

The last of the three most common transformations is translation. That is, a pixel shift of the image in the up–down or left–right axes. This translation can be of any size, but on a given night its size will probably be defined by the accuracy of the telescope drive you are employing.

The handling of such translations to ensure the alignment of image pairs is often a manual or semi-automated process. The user is required to identify a unique pair (or more) of stars on each image using a cursor and those coordinates provide the software with all the information it requires to scale, rotate and translate the images with respect to each other. Such a process can be undertaken by hand, but it can be a rather slow and frustrating affair.

When undertaking translations we must be aware of the fact that they may slightly degrade the quality of an image – in particular its resolution. Moving an image by a whole number of pixels is not a problem, whereas, moving an image by a fractional number of pixels means that interpolation is needed to work out the brightness of every pixel in the new image. Interpolation is a sophisticated technique for estimating the value of a given pixel based upon the brightness of the pixels that surround it. Similarly, rotating and scaling images will generate subtle effects that will leave the new images entirely acceptable visually but may affect the usefulness of the image for astrometry or photometry.

Image Stacking

This is a technique whereby several images of a scene are aligned with each other and a new image is created, combining the signals from each of the component images. This simple technique has several advantages. The first and most important is that it allows us to improve the signal-to-noise ratio (SNR) within the image.

Perhaps this needs a little discussion. Noise in images is a fact of life. No camera sensor yet built, or likely to be built, is perfect and any image taken will contain some noise arising from a combination of flaws in the read-out electronics and from heat within the sensor chip. As if this wasn't bad enough, light does not arrive from objects in a regular stream with a regular and consistent gap between each arrival. Instead it arrives in a stream with random gaps between each photon or even in clumps (a bit like buses). So, if you count all the photons that arrive within a certain time, you will find the total varies about an average value, with the range of values found being a measure of the noise of the image. In an astronomical image this noise manifests itself as slight pixel-to-pixel variations in the image brightness, and is most readily apparent against faint nebulosity. This source of noise should not be confused with that generated within the camera itself.

Mathematically, the SNR of an object is the brightness of a pixel divided by the brightness of the associated noise. So, crudely, if an object has a count of 2,000 and the noise count is 10 the SNR is 200 (2,000/10). The subject is more complicated than this, but that is broadly how it works and a good starting point for further thought. Howell's *Handbook of CCD Astronomy* listed in the "Further Reading" section contains a good discussion of image noise.

If we are attempting to image a very dim object, we will find that the pixel count caused by our object will not be much greater than the noise, i.e., the variation in brightness of the dark parts of the image. As you might expect, the whole scene will appear grainy, which is a sign of a low SNR. An image with a bright target and a flat dark background will signify a high SNR.

The problem we must address is: how can we remove the graininess caused by the noise and create better pictures? Using longer exposures to average out the noise is one approach, but there is another way. Take lots of pictures. When we combine lots of images of the same scene together, the noise component of one image tends to cancel out the noise in another and the final image can be very dramatically improved. This works because we have even numbers of pixels that are slightly too bright and slightly too dim because of the noise. The stacking works very well and improves with the number of images stacked. Here are some examples; if you stack four images, the noise can drop by a factor of 2, stack 25 images and it can drop by a factor of 5, stack 100 images the final image can have its noise reduced by a factor of 10. Those of you who like doing math puzzles will have recognized that the extent of the reduction achieved is related to the square root of the number of images stacked. This means that if you took 144 images the noise reduction factor would be the square root of 144 or 12 (as $12 \times 12 = 144$). Table 7.1 gives a list of potential results – what you would get in the ideal case only.

In practice, these reductions are only round figures and will not normally be achieved – especially if some of the noise in the images arises from electrical interference or is due to insufficient care taken in dark subtraction or flat-fielding.

The other great advantage of stacking images is that it allows us to overcome the limitations that are imposed by the effect of drive imperfections of telescope tracking. For example, using an old Vixen Polaris mount I find that the telescope drive varies slightly in speed over an 8 minute period during which a star will appear to drift cyclically by about 20 seconds of arc. Similar behavior will be found for all gear-driven telescopes with only the size, smoothness and period of the error

Table 7.1. The relationship between the number of images stacked and the improvement in SNR in the final image

Number of images	Improvement in SNR
4	2
9	3
25	5
64	8
100	10
500	22
1000	32

Figure 7.12. The impact of stacking different numbers of aligned images upon the variation in the sky background signal and the image SNR. The results show the result of stacking (left to right) 1, 9 and 25 images, respectively. Note the flatter background and more clearly defined stars. Image credit: Grant Privett.

depending on the manufacturer. By experimentation I have found that most exposures of 30s duration will be untrailed. As a rule I use at least 60 minutes worth of images for any deep sky picture, which means I must combine the signal from 120 untrailed images. Image stacking routines make this quite easy and in this instance would provide a SNR improvement of a factor of around 11.

A final reason for image stacking is that it provides the ability to combine multiple images that were taken on different nights, or even using different telescopes and cameras. Fine examples of such combinations are Nik Szymanek's images combining color images obtained using small amateur telescopes and cameras, in conjunction with imagery taken using telescopes from the Isaac Newton Telescope group on La Palma.

Another example shown (Fig. 7.15) demonstrates the technique taken to extremes: it shows the field of a dim quasar imaged 450 times over five nights. The images were stacked together yielding an image very much less well focused than an image obtained by the 1.2 m Schmidt camera at Mount Palomar, but with a similar sensitivity and level of noise. Not too bad for a 250 mm (10 inch) telescope from light-polluted England. It is fair to say that in each of the 30s images used the quasar was pretty much invisible. We truly are digging data out of the noise.

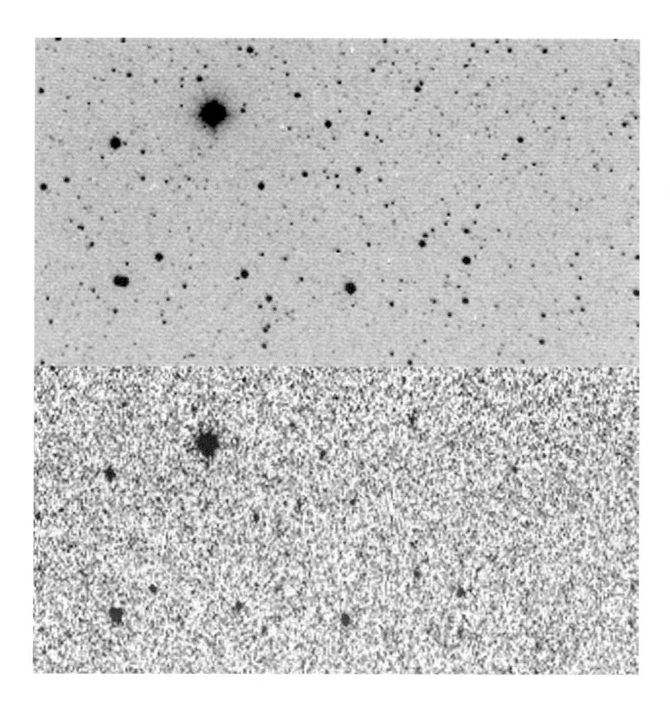

Figure 7.13. These images were each created from 10s of exposure time using the same camera lens. One is a single 10s exposure, the other the sum of 100×0.1 second exposures. The 100 exposures were carefully dark subtracted and flat-fielded. The moral is that there is a limit to what image stacking and co-adding can do. The limiting factor was the sensor readout noise. When the exposures become very short the bias/readout noise swamps the signal from the sky and even stacking cannot always save the result. Image credit: Grant Privett.

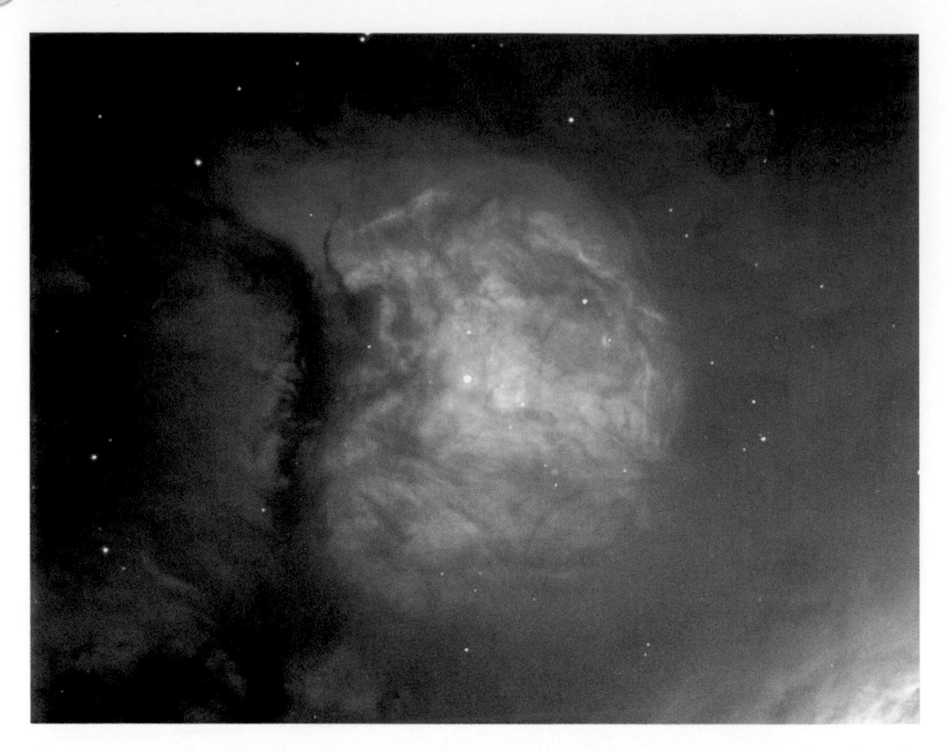

Figure 7.14. An example of color images from two separate sources aligned and combined. A monochrome image of M43 taken with the Jacobus Kapteyn Telescope (JKT) 1 m telescope on La Palma was colored using RGB images obtained using a commercial SCT and standard filters. Image credit: Szymanek and Tulloch.

As was mentioned earlier in Sect. 7.7, "Image Transformations", when aligning images through rotation or translation, pixel values are interpolated, which can lead to a rather odd sort of blurring or smearing of the background noise. This can be quite disconcerting to look at, but it is generally invisible once several images are stacked.

The method used to organize the aligning and stacking of the images varies from package to package, but they boil down to the same thing. One or two reference points are selected on a reference image and then all the other images are transformed/aligned so each point on the first image lines up with the same pixel in all the other images, thus overcoming the effects of the telescope drive errors. The commonly encountered options for this selection are discussed briefly in Sect. 7.10, "Automated Image Reduction". In the meantime, however, it remains worth considering how the contributions from each image are combined to create the final image. By now we are all familiar with the options of "Mean", "Median" and "Clipped" we discussed previously, which is just as well given that you will almost certainly find these will be offered in one form or another. One useful addition to those discussed earlier will be the "Sum" (or "Summattion") option. In this approach all the contributions to each pixel are added together. This can be very useful, especially if you intend to undertake photometry upon your images.

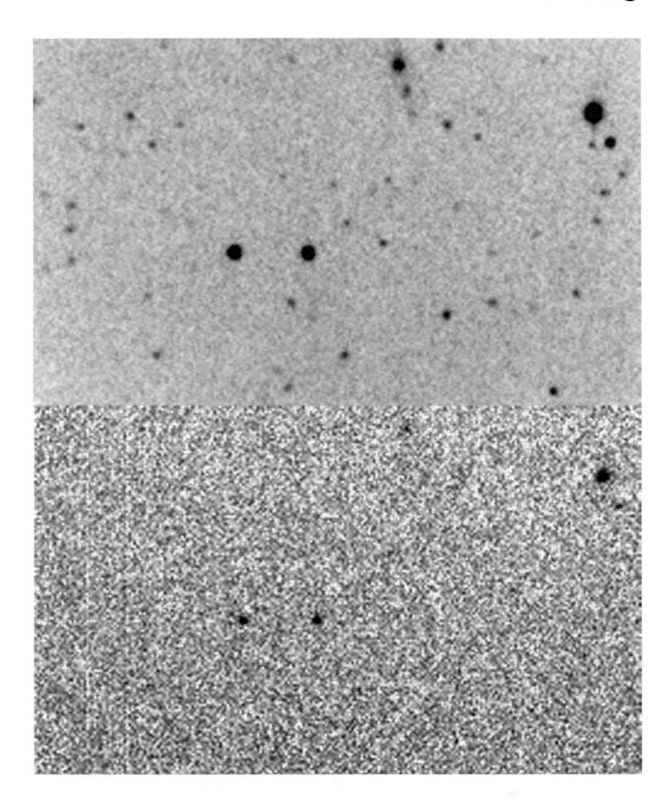

Figure 7.15. A field of view containing a $Z = 4.7$ quasar as imaged using a 250 mm (10 inch) Newtonian and a noisy homemade CCD camera and 30s exposures. The lower image shows the result from a single image, the upper image shows the result after combining the median stacked results of 450 images taken over five nights. The images were median stacked in batches of 50 and summed afterwards. Image credit: Grant Privett.

It is easy to overlook the fact that in summing images you may create images with a range of gray values far beyond the range of the sensor you utilized to generate your original image. Consider the case of two 16 bit (approximately 65,000 gray scales) images showing a star with a count value of 40,000. If two images are stacked together the star will now have a count of 80,000 which is beyond the range of simple 16 bit imagery. Stack several images and the sum can quickly be in the millions of counts. At that point it is simplest to store the data in a 32 bit FITS format and accept a larger file format that may not be readable by the software that came supplied with the camera you used to create the original images.

An alternative – if photometry is not being used – is to convert the image to a 16 bit representation by dividing all the pixel values by a constant factor, but it seems a shame to throw away information that we have worked so hard to gain.

The FITS format can store data of a variety of types and so is used universally within the astronomical community for storing anything from a spectrum, through normal imagery to multidimensional datacubes – a sequence of images varying either in wavelength or chronologically. By using this format you guarantee that almost any professional package will be able to access your data and does

Figure 7.16. Images created by stacking 10 images using three popular stacking methods (from top to bottom); mean, clipped mean and median stack, respectively. The images have been scaled similarly. Image credit: Grant Privett.

not take the risk of saving your data in a proprietary format that might, in just a few years, become redundant.

Image Mosaicing

Having created a well-processed image of a given small object such as M57 or NGC 6543, you might become a little more ambitious and decide to create, for example, an image of M35 plus its neighboring cluster NGC 2158 in the same field of view. If the field of view required is larger than the area of sky that can be imaged in one go by your camera/telescope combination, you will need to combine overlapping images into a mosaic of the surrounding sky. To do this you will need to use stars within the regions where the images overlap as the reference tie-points that define how the images are orientated relative to each other.

Unfortunately, whilst this process sounds simple in principle, it can, in practice, be quite difficult to do manually. Fortunately, software can now help us out a lot, but it is worth understanding what is going on. Generally, the problem is tackled in two simple stages.

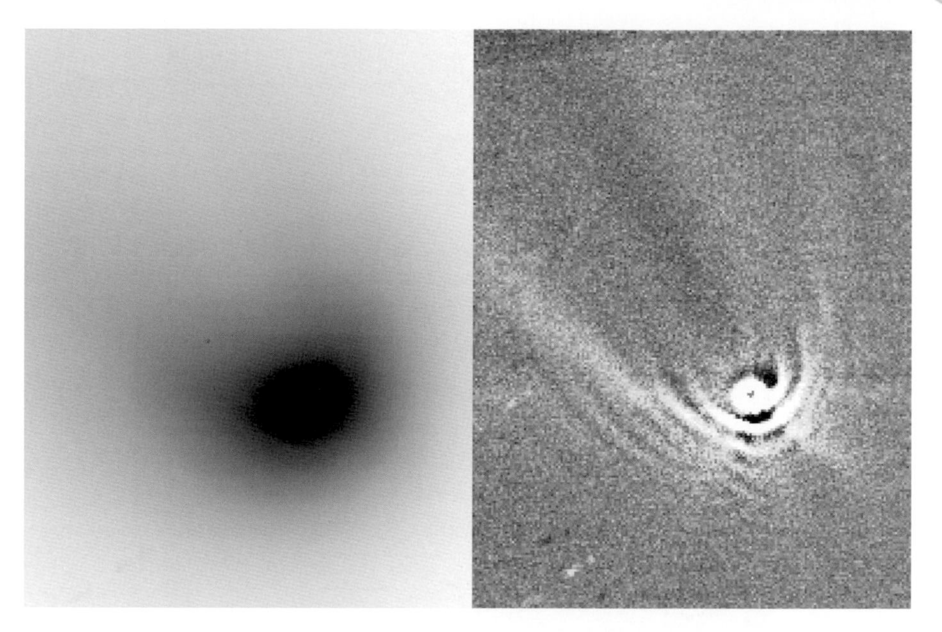

Figure 7.17. An image of the inner coma of comet Hale–Bopp on March 30th, 1997. At this point the rotating core was spewing out jets of gas and dust that appeared to wrap around the core forming shells. Note that by the time this picture was taken the comet was moving so fast across the sky that median stacking images using the core as a reference point meant the background stars effectively vanished from the image. Had the images been summed the stars would have appeared as parallel lines. Image credit: Grant Privett.

For simplicity we will consider an example where nine images have been taken so that they form a 3×3 image rectangle around the extended target you wanted to capture. Given the overlap between the images, the mosaiced image will be of the order of 25% smaller than three times the width and height of a single image. It is worth remembering that, for those of us who do not use computer controlled telescopes, just creating the image set itself can be quite an accomplishment.

In the first stage, the software looks at how the images will overlap and identifies the stars that are shared by the central image of the nine, and each of the surrounding images, in turn. It then examines each of the overlap region stars to determine its peak brightness compared to the brightness of the sky background pixel count. The information derived provides a factor by which all the pixels in the outer image are multiplied to keep the stars of consistent brightness from image to image. In some systems you may be asked to identify the tie-point stars on each image using the cursor.

You might at this point be wondering why the brightness of a star might vary so much between images and also why the sky brightness would not be the same everywhere on the sky. The answer is that a star will appear to grow brighter or fainter during the night depending on whether it is rising or falling and can, in addition, brighten or fade if the sky transparency varies due to haze or mist. As this can take place at any time, the brightness scaling factor values can change quite a lot in only a few minutes.

Figure 7.18. A lunar mosaic painstakingly created by Simon Clayton-Jones in 1995 by aligning 50 images taken with the monochrome Connectix *Windows 3.0* camera of that year. This labor of love demonstrates how even small images can be combined to create an image of greater sky coverage. The dark areas occur where no images were collected. Image credit: Simon Clayton-Jones.

The other important adjustment made ensures that the sky background pixel counts of all the images are adjusted so that they are all the same. If this is not done, then the resultant mosaic would show clear demarcations – a sudden change in the sky brightness – where the images joined up. Good image processing software avoids this unsightly effect by normalizing the background of every image so that all images are of equal brightness. Despite this, it would be wise to avoid

Figure 7.19. A number of diverse images – including some from the HST – of the M101 spiral galaxy allowed the construction of a mosaic 12,000 × 16,000 pixels in size. This monumental effort is one of the most detailed images of the galaxy ever produced and at full resolution the file is over 455 Mb in size. Background galaxies are clearly seen (inset) and the spiral arms resolved sufficiently to show the brighter blue giant stars individually. Image credit: NASA and ESA.

creating mosaics when the Moon is nearby. Similarly, if it is a moonless night, avoid imaging low-lying objects as these will almost certainly have backgrounds containing brightness gradients caused by glow from the sky near the horizon.

Having corrected the sky background values and adjusted the brightness of each image, the software uses the overlap star positions to align images to each other. This simple image mosaic should have no major artifacts but it may be found that there is more to be done manually, such as cropping the image before moving on to image enhancement.

To a greater or lesser extent *AstroArt* or *Maxim-DL* and *Registar* handle image mosaicing automatically. *Photoshop* allows the display of each individual image in such a way as to hide any differences in sky background by employing the "Curves" function.

Alignment can be done very accurately and usually works very well for images taken on the same night, where the image orientation is likely to be the same. But for images taken on different nights, when the images were rotated relative to each other, it can be less successful. Very large fields of view are particularly difficult and take a lot of practice.

Figure 7.20. When images are aligned and added – rather than aligned and median stacked – the sky background signal means the edges of the images are fainter because less images contributed to those pixels. The image shown has been scaled to show how aligning and adding is not a method suitable for mosaicing. Image credit: Grant Privett.

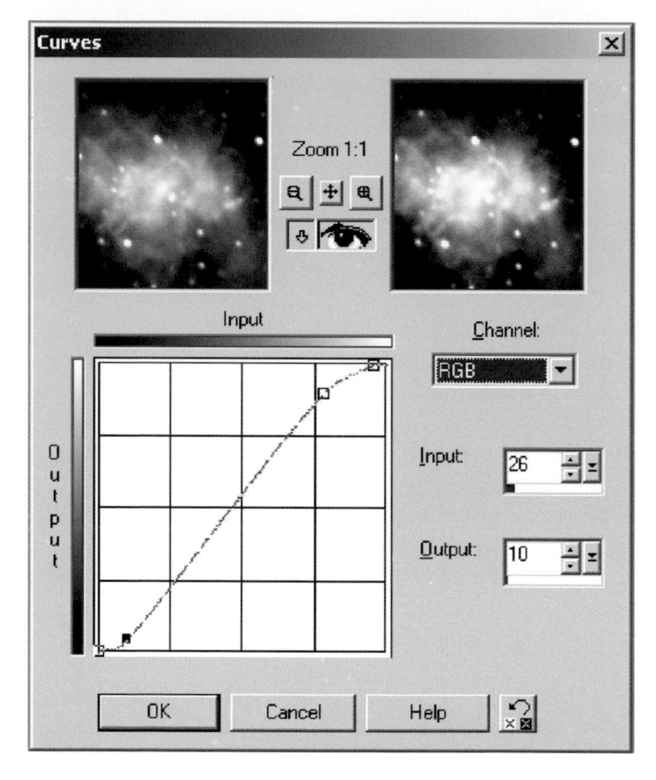

Figure 7.21. An example of the "Curves" utility – in this instance from *Paintshop Pro* – for image enhancement. By bending the line you affect the transfer function and how the image will look. Image credit: Grant Privett.

Automated Image Reduction

As we have seen, flat-fielding and dark subtraction are essential processes in creating the best quality astro-images, but by now you may, understandably, be getting the impression that image reduction is a long and rather tedious process. Last time I tried manually processing 30 or more individual images it was a very dull way to spend an evening and you could be forgiven for looking for shortcuts. Fortunately, that is exactly what most dedicated astronomical image processing packages provide.

For example, *AstroArt* provides its "Preprocessing" option, *CCDSoft's* "Image Reduction" option and *AIP$_4$Win* via its "Calibration" options. Each of these, in its own way, provides a means of telling the software which files are dark frames, bias frames or flat-fields. The file selection is usually achieved using the "drag and drop" manoeuvre familiar to all Microsoft *Windows* users. That is, a list of files is provided, together with the ability to browse round the file/directory/folder structure as necessary, and files are highlighted using the shift key or a mouse and then dragged into the appropriate screen location. One useful feature of these packages is that most do not force you to use a fixed approach to imaging. They do not require you to have darks, flats or bias frames, but instead facilitate the use of those images you do have.

In addition to identifying the files that will be involved, they ask the user to indicate the method that will be used to create their master dark/flats/bias frames

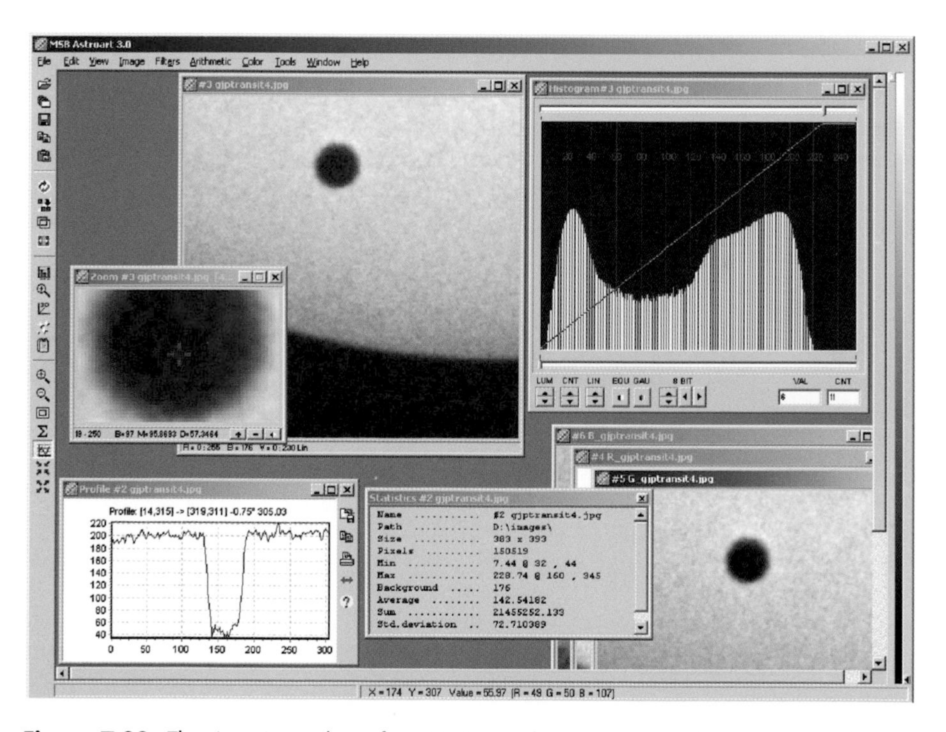

Figure 7.22. The *AstroArt* package from MSB provides a quite powerful and reasonably priced image processing and image acquisition package. Image credit: Grant Privett.

(by median or mean, etc.). This might sound a lot to set up but usually the default setting can be adopted by the uncertain. This is very useful as it means you can quite quickly try two or three different options on a given set of images and see which works the best. As an example, this afternoon I processed an image set consisting of 100 images of 500,000 pixels each. These were processed together with 20 darks, 10 flats, 10 bias frames and 10 dark frames for the flats; the whole process running on a 3 GHz PC took rather less time than it does for a kettle to boil. That's an incredible advance on the early years of amateur image processing when reducing an image set manually could take a whole afternoon or in some cases a whole day – even cheap machines are very fast these days.

The other option the software packages allow you to select is the manner in which the images are to be aligned. These will generally come in several possible forms:

1. Manual intervention to indicate a reference star (or two) on the first frame and to subsequently confirm that the software has successfully located the same star on all images. The two star identification variant will be required for image sets where the camera was rotated or removed at some point during the evening – as can happen by accident or if returning to the object on a second night – or if the images come from different cameras.
2. An automated search for the brightest part of the image which is then used as a reference – which can cause problems if the brightest part of the image was right at its edge and moves off during some exposures, due to drive inaccuracies.
3. A planetary alignment system where the planet and its center are located and then all subsequent frames aligned to the first frame, using correlation techniques.

All these options and any others provided are worth experimenting with, as the time spent will be rapidly repaid as your experience grows. Don't be surprised if your first few evenings of image reduction seem painfully slow, confusing and unproductive. As your experience and confidence grow the available options will become familiar.

Some packages even go so far as to provide the option of automating some of the image enhancement processing. I cannot say I am a big fan of this concept as the creation of a fully reduced and stacked image should initially precipitate only a close examination of the resultant image. The examination should look for serious flaws. For example, are the star images circular, are there obvious processing artifacts present, is an obvious sky gradient present and have you detected the object you set out to image, an issue familiar to those looking for faint diffuse targets without a computerized telescope mount? The creation of the fully reduced image may even be the appropriate time to determine whether or not the image quality is good enough to warrant keeping it, or can signify that another imaging session may be called for! Certainly, I would not be inclined to modify the fully reduced image until I had made sure it had been saved on disk.

Given the amount of time you will save by using batch/preprocessing of your images, it would almost seem deliberately perverse to ignore them. It is worth playing with a few simple image sets to examine the intermediate stages but, in general, it is not a great way to spend your spare time.

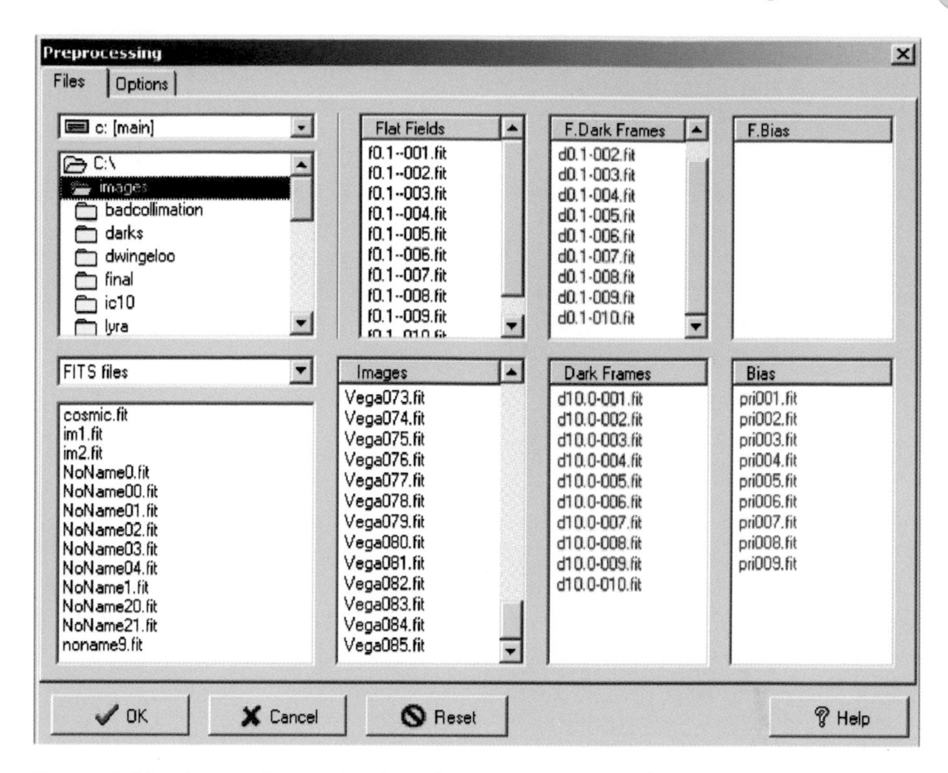

Figure 7.23. The use of a graphical interface to select and identify files as darks or flats, etc., has greatly simplified the automated the processing of images. Image credit: Grant Privett.

Image Handling

It's worth considering some of the other problems that can occur when handling images. Until recently, the problems of the image processing fraternity have not been a concern of most PC users. But the rapid growth of the digital imaging market arising from webcams, digital video cameras and mobile phones is now having an obvious impact upon the marketplace, so you increasingly see computer specifications targeted at the home user who is image processing. That's good news for us.

There was a time, not that long ago, when storing images was a major concern. In 1992 for example – when I first started writing image processing software for my own camera – a 25 Mb disk and a 20 MHz CPU were enviable. Unfortunately, even then, image sizes were measured in the hundreds of kb so a full hard disk might contain only a hundred or so images – allowing for the operating system. As a consequence, if you allow for flats and darks, you might only have space for 50 or so images on the disk before you needed to archive some of them to a floppy disk or, even slower – and more expensive – tape. Happily, hard disks have grown enormously and are now much bigger, with 120 Gb being the current norm, which means that even with a 16 bit 10 Mb pixel camera there is space to store over 6,000 images.

The archiving situation is also greatly improved. At one time, storing all the images for a given object might require a stack of floppy disks, but now, by storing the images on a writeable DVD, it is possible to keep several nights' images on media costing around 60p ($1). With a price that low, especially for those of us blessed with few clear nights, it's well worth permanently keeping all your images, in case you decide to reprocess them at a later date, when you have grown in experience or a new technique has become available.

As disk storage of any type is now rather less of an issue than it once was, the most important issues for image processing have become:

- the speed at which your machine can read a disk;
- the rate at which it can read images stored in memory;
- the amount of memory that your machine has.

Obviously, expense is a great concern to all of us, but where possible the simple advice to give would be:

1. Ensure that your machine has at least twice the minimum memory specified for your operating system. For *Windows* 2000 this amounted to 512 kb, while for *Windows* XP 1 Gb would make more sense. Less will work, but everything may take longer and involve a lot of temporary caching to hard disk.
2. If possible use a disk that is SATA compliant and of 7,200 rpm speed or better.

   This will ensure that the time taken to convey the image information from its disk stored form into the software is reduced to the minimum. If, for example, you are processing 120 images and it takes 2 seconds to read each image then four minutes will have elapsed before the software even starts to think of doing anything useful to the data.

3. Use the fastest memory that your computer motherboard supports.

   This will help ensure that when the software manipulates the information stored in memory it can read it with the minimum possible delay.

   This will also ensure that the machine has plenty of work space within which to handle images and will not need to temporarily write them to disk if you start to use other software while the image processing is still underway. One of the reasons for worrying about memory so much is that, while your images may take (say) 4 Mb of space when stored on disk as integers (whole numbers), once they have been read into the image processing software, they may be converted to a real (numbers including floating points) format and then take up 5–8 times as much memory space. A single 2 Mb image might become 32 Mb which would mean that storing only eight images in memory would be enough to completely use up 256 Mb. That's not many images.

4. Employ the fastest CPU you can.

   Currently, manufacturers of chips are still managing to keep up with the well-known Moore's law (crudely put, computer CPU processing power will increase by a factor of 2 every 18 months or so) and so PCs are continuing to rapidly increase in speed. This means that occasionally upgrading your processor/motherboard is a worthwhile thing to do even if much of the rest of your computer stays the same.

Given that performing a maximum entropy deconvolution or even a simple unsharp mask upon a large image can require billions of calculations, it is worth getting a quick machine if you possibly can. That said, it should never be forgotten that the best place to invest money is in your telescope, camera and mount.

The final issue might be the format in which you store images. There are several possibilities but the easiest thing to do is to translate your images to FITS format – if they are not already – and keep all the results in the same format. When you get an image to the stage where you want to share it with other people you can store it as one of the more commonly encountered image formats such as TIFF, BMP or JPG. Remember though that JPG is a compressed and lossy (i.e. some subtle detail will be irretrievably lost during compression) format that should only be used in its very highest quality mode. The compression can be quite significant, which is good for displaying images on a website or for emailing them.

All in all, things are rather easier than they once were. Most modern home PCs are equivalent in power to the supercomputers of the early 1990s and whilst I do upgrade my PC, I do so not because the software is too slow but because of operating system and peripheral issues. Today's machines are so fast that very few image processing operations take longer than the time needed to make a cup of coffee and I only image process on cloudy nights!

Histograms

An image histogram is, at its simplest, a way of showing graphically how well exposed an image was. The histogram graph normally shows the pixel count range of your sensor along the *x*-axis (left-right), while the *y*-axis (up-down) shows how many pixels there were with each of the possible pixel count values.

As an example; if you look at the histogram of an image of an open star cluster with a 16 bit (pixel count range 0–65,535) camera and there are no pixels brighter than 32,000, then it is clear that the exposure could have been twice as long – telescope drive accuracy permitting. Similarly, if the histogram instead indicated that the image had more than one or two pixels with a brightness value of 65,535, then some of the pixels may have been over-exposed. In such a case, reduce the exposure length slowly until no pixels are saturated.

Now, you will have realized that there is a potential problem with this approach. Most of the pixels within the image will be of a count value that is very close to the sky count value. This will mean that if the histogram attempted to display a peak showing (say) a million pixels due to the sky count, while no other brightness level was present in more than a few hundred pixels, the rest of the graph would be impossible to discern. It would be a near flat line followed by a sharp peak and then a quick decline to the flat line once more. To overcome this, you will find that the *y*-axis of a histogram is usually scaled logarithmically. For those of you who have not encountered logarithms before, they are a very useful area of math that helped to speed up calculations in the days before handheld calculators became widely available, and slide-rules were the only alternative. It is easiest to say that on a logarithmic graph, the difference in spacing between a *y* value of 100 pixels and 1000 pixels will be the same as that between 1000 pixels and 10,000 pixels.

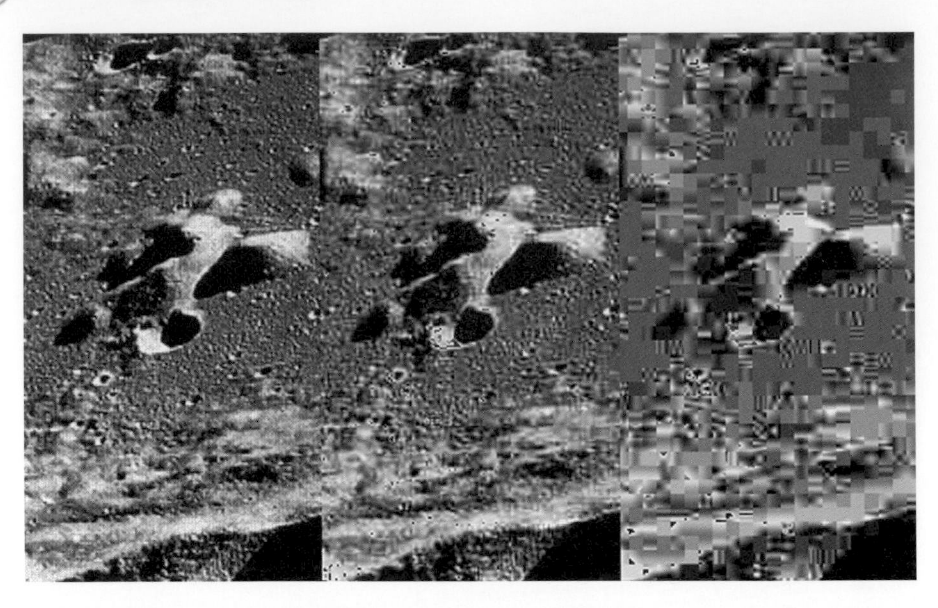

Figure 7.24. This image shows the impact of image compression upon the quality of an Apollo 11 photograph. As the image is compressed square artifacts become apparent. The left-hand image is uncompressed, the middle image is compressed by a factor of 10 and the right-hand image by a factor of 20. Image credit: NASA.

Now that may sound a bit odd, but it means that using a single histogram graph you can see – with tolerable accuracy – that there were only 2–3 pixels of a particular count value, at the same time as seeing there were roughly 100,000 at another count value. In effect, the graph scaling is compressed, making it easier to read.

Image histograms have another use that you will frequently encounter. When software has to display an image without any prior information from the user as to which pixel count should be displayed as black and which as white, we would ideally like it to show the image in such a way that everything on the image can be immediately seen. Achieving that is a lot trickier than it sounds. If you display the image with black as the pixels with count value 0 and (for a 16 bit camera) every pixel with a value of 65,535 as white you will normally get a pretty awful result. Similarly, just using the count value of the dimmest pixel as black and the count value of the brightest pixel in the image as white will give disappointing results. Those are often presented as possibilities in the pull down menus but are rarely more useful than automated methods.

The rules that are most frequently applied to automatically scale the brightness/contrast of a display image are a compromise based upon the image histogram. One such rule is the 5–99 percentile rule in which the software initially determines the pixel count value that will be brighter than the darkest 5% of the pixels – which it then uses to define its "black" value. The histogram is then re-examined to determine the pixel count value that is brighter than 99% of the image pixels – which it uses to define a "white" level. Every pixel of a value within these limits is displayed as a shade of gray. In many cases – mainly for deep sky imaging – this rule will give

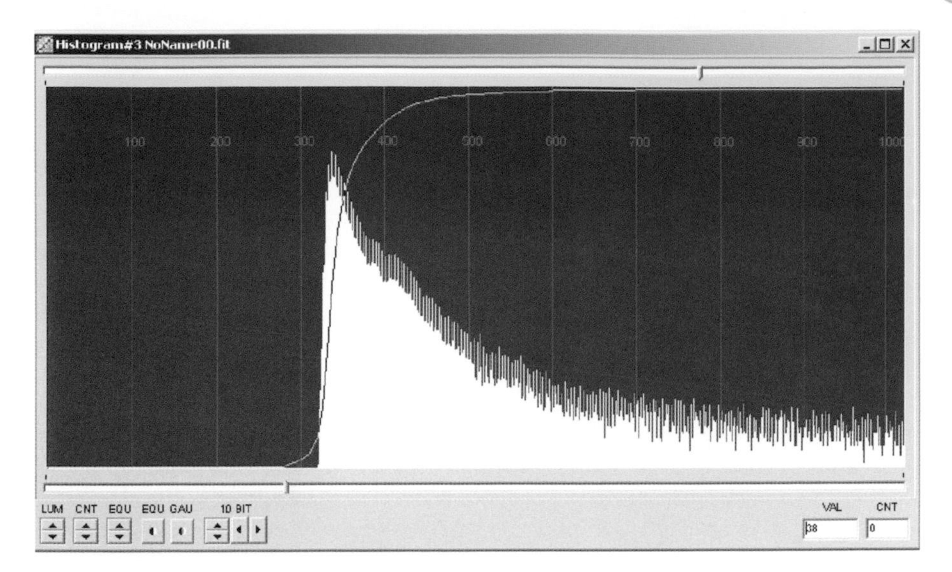

Figure 7.25. The histogram of an image containing substantial nebulosity. The histogram of a sparsely occupied bit of sky would have a slightly sharper histogram peak. In this instance the *y*-axis is displayed logarithmically. Image credit: Grant Privett.

a good result, but it will often fall down on images of planets where a 90–100% rule (or variant) will be much more successful. This is because the detail you are interested in bringing out is really a subtle difference in brightness between adjacent bright pixels rather than bright pixels against a dark background.

Most software packages will offer you the choice of how data is to be scaled, but you will find that those packages specializing in the processing of webcam imagery will usually default to a rule applicable to planetary images. As before, experiment with the automated methods available. Find the option that gives the most satisfactory result and then tweak the black and white limits manually – this may be implemented using sliders, toolbar buttons, buttons on the histogram display or two of the function (F1–F12) keys on the top edge of your keyboard. Every program seems to do it differently. The next section describes how once the black and white limits have been chosen, the display of all the values in between can be optimized.

Brightness Scaling

When looking at images, even those where you have carefully chosen the pixel count values that should correspond to white and black, you may find that it remains difficult to see the subtle detail. Take, for example, a spiral galaxy image. Frequently, the center of the galaxy will be quite bright compared to the spiral arms, so when we adopt the brightness of the core as defining white the spiral arms will be a dull near-featureless gray – not what we want from a galaxy's most attractive feature. One potential response would be to just lower the pixel count value that corresponds to white, but that would result in a large part of the central portion of the galaxy becoming a burnt-out brilliant uniform white well before the spiral arms were clearly seen.

You have probably seen something similar in normal photography when trying to see detail on the faces of people photographed when backlit by the Sun. All the detail is in the image, with the scaling making the sky appear very bright but the contrast is so severe that any detail on the faces will be lost.

To overcome this serious limitation of normal – i.e. linear – scaling between black and white, a number of other scaling types are offered by image processing software. The scalings, which are also known as transfer functions, most frequently encountered are logarithmic, exponential, square root, power law and a diverse range of histogram equalizations. As you can see, all these have their origins in some type of mathematical function, but their origin is not an issue. Unless you are working at an advanced level they are merely tools to be used like any other. Most people do not know the detail of what happens after they press the ON button of their mobile phones, but that has not limited their popularity. It's the result that is important. On the other hand, if you become curious and want to see exactly what's going on behind the GUI, I would encourage you to have a look at the "General Image Processing" references in the "Further Reading" section, which have been chosen for their clarity and accessibility.

Figure 7.26. Using a linear scaling to allow the spiral arms of the galaxy to be seen means the central portion of the galaxy is displayed as white and the detail hidden. Image credit: Grant Privett.

They are all useful, with some functions being designed with a very specific type of target in mind. For example, the Gaussian histogram equalization finds its greatest use when applied to galaxies. In essence they work by modifying the way we set the shades of gray that exist between black and white. Consider an image where the black pixels have a pixel count value of 0 and the white pixels a count of 1000. In the linear scaling method we are most familiar with, image pixels with a count of 500 will be colored a shade of gray exactly half way between black and white. But, in the other scaling types that will not be the case and as the pixel count advances from 0 to 1000 the shade of gray will change either more quickly or more slowly than you would encounter for a linear scaling. So a mid-gray pixel of value 500 might appear nearly white in logarithmic scaling, while looking nearly black when power law or exponential scaling is instead applied.

The rate of the change from black to gray is defined by a relatively simple – depending on your viewpoint – mathematical function. The function provides a smooth transition from the black to gray. The rate of the change is in some simple cases defined by a "gamma" value where a high value means a stronger contrast while a lower value means a softer appearance. Photographers will be familiar with this idea having encountered it in the "hardness" value of a photographic print paper. Logarithmic or even histogram equalization scaling combined with

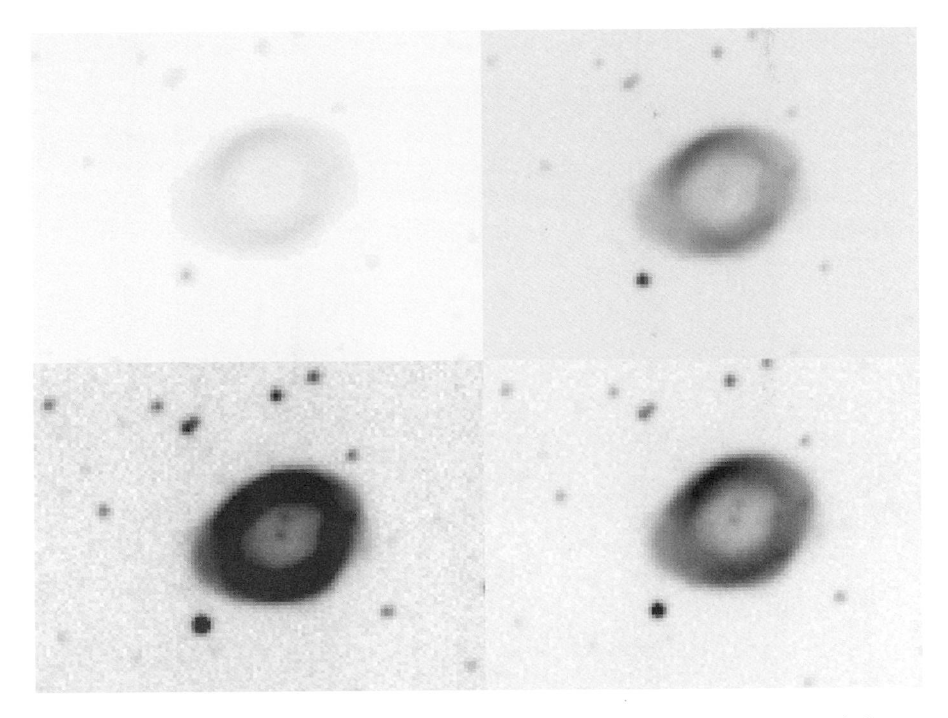

Figure 7.27. Four examples of the linear scaling of an image of M57 – The Ring Nebula. From top left clockwise; scaled to the full range of the sensor, scaled to the data range of the full image, scaled from the sky background to the brightest portion of the nebula, scaled using the 8% and 99% histogram values. Image credit: Grant Privett.

carefully chosen black and white levels is particularly useful for ferreting out subtle detail that might be missed. Used judiciously, you should be able to generate good results even if working on pictures that are the equivalent of a mole in a coal cellar, or a polar bear in a snowdrift.

Histogram equalizations are the exception to the idea of a smooth transition from black to white. Equalization techniques examine the image histogram and attempt to optimize the transfer function to ensure that the maximum possible detail is visible. It does this by making sure that if most the pixels contained values near to (say) 200, then most of the gray shades are used to display pixels that have a count value of near 200. This is great for looking at very faint and diffuse objects that might otherwise go totally unseen. Even for obvious objects it can reveal the full extent of an object. This latter ability is especially useful with globular clusters and diffuse nebulosity.

Unfortunately, particularly if the black level is carelessly chosen, a histogram equalization can have a very harsh effect upon an image, producing ugly and grainy representation. Because of this you will often find it available in a modified version that softens the result.

Rather than attempting to describe the mathematics behind these scaling/transfer function options, it is best that you try applying each of them in turn for every object that you image, and see which produces the best result in each case. If you already have some images, try the scalings out on a variety of objects including

Figure 7.28. Four images of the globular cluster M15. The images are scaled between the same black/white thresholds using different scalings/transfer functions. The examples shown are, from top left clockwise: linear, logarithmic, histogram equalization and DDP.
Image credit: Grant Privett.

globular clusters, galaxies and nebulas. You will find that the result will vary somewhat with the black and white levels chosen, so there is a lot of scope for experimentation. Ultimately, you will find that there is a combination you generally use for a quick look-see. As you can see, when we abandoned a linear scaling we entered a world where the best thing to do to an image starts to become a matter of opinion rather than fact.

It could easily be argued that image brightness scaling is actually the first stage of image enhancement, but this only really becomes the case if you modify the pixel values in the source image. Be sure to take care that you only modify how an image is displayed and not the pixel values themselves. Be sure to understand the distinction. There is still some astronomical image processing software available that will modify the actual pixel values so that all pixels below a certain black value have the same value. At the same time it will find all the pixels above a higher white limit and set those to the same value. This could conceivably be useful, but more often than not it is bad news. If you are not careful, the information you worked so hard to obtain could be destroyed and you cannot go back without opening the file again. What if you change your mind about the image and decide the black limit should have been a tad lower? Merely defining how an image is displayed rather than changing pixel values is by far the safest and least frustrating approach.

False Color Rendition

We have described how the information in an image can be displayed using a brightness scaling/transfer function that helps to reveal subtle details that would otherwise be lost due to a lack of contrast. Another way to approach this issue is by assigning colors to each of the brightness levels and varying the color as you

Figure 7.29. An image of IC10 where the two halves have been scaled differently, on the left using a logarithmic scaling and on the right a histogram equalization. Image credit: Grant Privett.

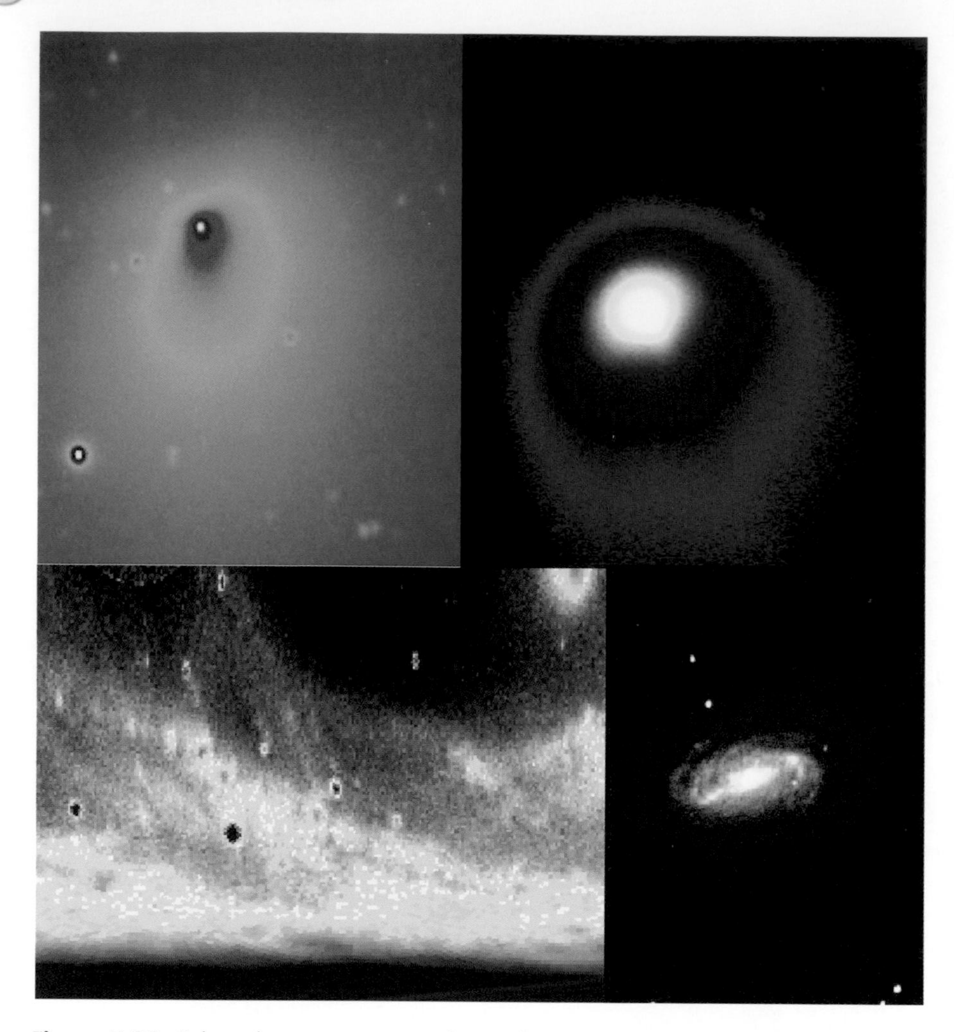

Figure 7.30. False color representations of monochrome images using a variety of color palettes – some more tasteful than others. From top left clockwise they show Comet Hale–Bopp in June 1996, Comet Hyakutake in March 1995, NGC 2903 in Leo and a polar representation of M51, the Whirlpool Galaxy. Image credit: Grant Privett.

progress from what would have been black through to white. This color sequence forms a palette that is used to draw the image.

As you might imagine, there are hundred of ways in which this could be done; you might for example run through red, orange, yellow, green, blue to end at indigo – the sequence found in the rainbow – but in practice only a few are widely used. The appearance of individual palettes is probably best demonstrated visually, but popular types of variation include:

Black > gray > white (normal)
White > gray > black (negative)

Black > red > yellow > white
Black > blue > green > yellow > red > white.

As there are many other possibilities you can expect to find at least five provided by every astronomical image processing package. This may not seem a large selection given the range possible, but when these color palettes are combined with the brightness scaling transfer function it's a very powerful capability.

The colors they use to illustrate an image are entirely false – I for one have yet to see a bright green galaxy – and have more in common with Warhol's oddly colored pictures of Marilyn Monroe than they do with the real color present in a scene. But it remains a very powerful technique even if it did become over-used in the early 1990s – before real color imaging became widely available within the amateur astronomy community.

There are a few other possibilities you are likely to encounter, the best-known of which is generally called something similar to "Zig-zag", "Zebra" or "Sawtooth". In this palette the color goes from black to white several times over the complete brightness range – the actual number of times being a user input. This provides an image that can appear to be semicontoured or even virtually incomprehensible – especially if too many cycles are selected. It is only really of significant use when very subtle differences in brightness are being looked for, and I would probably use some sort of histogram equalization instead.

CHAPTER EIGHT

Other Techniques

You will find that there are many options in the menus of astronomical image processing packages. Not all could, strictly, be classified under the heading image processing, as they instead help to handle the image, while some provide information that can be useful.

Others amount to fudging the data somewhat to give a more satisfying overall impression. These might be described as image enhancement, but they are a little too *ad hoc*, being applied only to a small part of the image and requiring a lot of effort on the part of the user.

A final group displays the image in a different way that can be interesting, attractive or revealing.

Statistics

This will allow you to look at statistical data, relating either to the whole image or to a small part chosen using your mouse. It may provide information on quantities such as the mean/average pixel count, the sky background count value (usually the value found most in the histogram), the lowest pixel count value, the highest value, standard deviation of the pixel count values and the mode and median values. These values can be useful for looking at how successful dark subtraction has been, and in determining the amplitude of the noise within an image.

In the case of the lowest and highest count value, i.e. the faintest and brightest pixels, the software generally identifies where the pixels are within an image. The coordinates will be defined either in terms of the traditional Cartesian coordinates (x, y) so that going from the bottom of the image to its top increases the y-coordinate value, or in a form commonly found in image processing where moving down the image from the top increases the y value. In both cases going left to right increases the *x*-coordinate value. So, if stipulating the position of a pixel within an image, be sure to check whether the coordinates (0,0) occur in the bottom left or top left corner of the image.

FITS Editor

The FITS format is a file format used throughout the astronomical community. It stores all the image data plus information about the image such as its width, height, and whether the image is 8 bit, 16 bit or some type of floating point number. In addition, its text header section can contain a large amount of information about the image, such as the coordinates of the field of view, the temperature of the sensor, the length of exposure, the type of telescope, etc.

On a professional image this can run to dozens of labeled fields plus a number of comment fields. Most software lets you see the data in these fields, a few will let you edit them. So, you can add your own comments such as the site from which

```
BITPIX  =                   32 /       Updated by AstroArt
NAXIS   =                    2
NAXIS1  =                  858
NAXIS2  =                  889
EXTEND  =                    T / FITS dataset may contain extensions
COMMENT   FITS (Flexible Image Transport System) format is defined in 'Astronomy
COMMENT   and Astrophysics', volume 376, page 359; bibcode: 2001A&A...376..359H
LBOUND1 =                   58 / Pixel origin along axis 1
LBOUND2 =                   29 / Pixel origin along axis 2
OBJECT  = 'ESP - FASTMED Image' / Title of the dataset
DATE    = '2005-11-25T18:12:32' / file creation date (YYYY-MM-DDThh:mm:ss UT)
ORIGIN  = 'ING La Palma'                /Tape writing institution
HDUCLAS1= 'NDF     '               / Starlink NDF (hierarchical n-dim format)
HDUCLAS2= 'DATA    '               / Array component subclass

PACKTYPE= 'OBSVATON'                    /Packet type
PACKVERS=                    2          /Packet Version Number
PACKDATE= '96/06/24'                    /Date of packet creation
PACKTIME= '21:58:40'                    /Time of packet creation
PACKNAME= 'JOB49819'                    /Packet name
PACKPDAT= '96/06/24'                    /Date of previous packet of this type
PACKPTIM= '21:50:44'                    /Time of previous packet of this type
PACKPNAM= 'JOB49818'                    /Name of previous packet of this type
OBSERVER= 'GJP     '                    /Name of the Observer
TELESCOP= 'JKT     '                    /Name of the Telescope
INSTRUME= 'AGBX    '                    /Instrument configuration
OBSTYP  =                   40          /Observation type
RA      = '14:00:30.96'                 /RA of the source
DEC     = '+04:05:10.8'                 /Declination of the source
EQUINOX = 'J2000.0'                     /Equinox of coordinate system
ZENDIST =                28.49          /Zenith distance
AIRMASS =              1.13745          /Air mass
DATE-OBS= '24/06/96'                    /Date of the observation
```

Figure 8.1. An example of the data content of a FITS file header. Note that only a few of the fields are compulsory, but NAXIS and BITPIX are essential. The FITS standard allows for many variations and extensions making the format extremely successful. The example shown is for the JKT on La Palma and the image was taken in June 1996. Image credit: Grant Privett.

the image was taken, who took it and what the target was! This can prove very useful as a source of reference long after the night is forgotten.

Pixel Editing

This is a common feature that allows you to edit an image at the pixel level. A purist might see it as cheating, but the odd tweak can be helpful in removing a small obtrusive blemish. You select a part of the image using the mouse and the count values for a grid of pixels around the chosen location (or for a single pixel) are displayed. You can then type in new values for the pixel(s) that suffers from unsightly noise artifact.

This approach is especially useful when an old dark frame or flat has been used to reduce data. It will also prove its worth when removing solitary cosmic ray strikes without applying a global filter to the image. After such modification an image should not be used for photometry or astrometry – of which more later.

Bloom and Fault Correction

Blooming occurs when a bright object such as a star saturates the sensor – that is to say a pixel is asked to collect the signal from more photons than it has capacity to store. When this happens, the signal can overflow into the surrounding pixels or into the column the star center occupies. The net result is that a slightly bloated star image is seen, but in addition to that there is a narrow and bright feature leading up or down (or both) from the star. Many cameras have antiblooming hardware that minimizes the chances of these blooming lines (sorry) appearing,

Figure 8.2. The correction of cosmetic faults in imagery arising, in this instance, from an uncorrected hot pixel that meandered across the stacked image showing the shape of the periodic error of the drive. In this instance pixel editing was applied to minimize the impact, but it would have been equally acceptable to copy another part of the image background and place it across the offending blemish. Life gets more difficult in crowded Milky Way images. Image credit: Grant Privett.

but if the star is grossly over-exposed they will occur. They are frequently encountered when trying to image targets like NGC 7023 or M42 where a nebula has within it a relatively bright star(s). If the nebula is to be properly exposed it may be that far too much signal will be collected from the star and blooming will arise.

As the sharp lines hide what might really have been in those columns, the software cannot very easily replace what is hidden, but it can make a guess at what might have been there by looking at the columns on either side of the blooming line. The user marks out the area containing the streak using the mouse, the software samples the neighboring columns, calculates the mean value, applies a little noise (the same amount as it finds in the image) and overwrites the blooming line. This can be a very successful approach – especially on large images, but it can have problems on some types of nebulas, in very crowded star fields and if there is a bright star within the bloomed column. The details of the interface application vary, but generally an estimate of the background sky is employed.

Other more general tools may be encountered, the most obvious examples being menu options that can correct a "dead" column or row of pixels. All the user has to do is identify the column or row – whether by numeric input or by using the mouse.

Figure 8.3. In this over-exposed star image the four diffraction spikes induced by the spiders supporting secondary mirror can be readily seen. The unsymmetrical spike at the bottom of the star is caused by blooming. Image credit: Gordon Rogers.

Night Visualization

Many of the software packages that we use to reduce data also have the capability of controlling a telescope, CCD or computerized focusing mount – in some cases all at the same time. As might be guessed, this means the software can be used outdoors at night. Perhaps unsurprisingly, at night a laptop screen is blindingly bright. Even when viewed through three layers of dark blue OHP acrylic sheet the glow from a c. 2002 (not exactly state of the art) Dell laptop is bright enough to damage night adaptation. To overcome this, a night visualization mode is often made available. This changes the color scheme in use, so that menus, screen backdrops and everything else are a shade of dim red. This can be very effective, though some color schemes will still be far too bright for people wanting to watch for meteors while waiting for their camera to take pictures. The action is usually on a toggle basis, i.e. click it once for turning it on, click it again for turning it off. Amateur astronomers may be the only people moaning that their laptop screens are too bright.

Profiles

A profile is a graph showing the brightness of every pixel that lies between two locations on an image. Once you have selected two points (most frequently accomplished with a mouse but one or two packages still ask for coordinates to be typed in) the package finds all the pixels in between, examines their values and then automatically scales the data to fit the space available for the graph. So, if you draw a line between two points in the image background and the line does not cross any stars, the graph will show a near straight and level line with some variation showing the noise within the image.

You would think profiles of galaxies would be very interesting, but in truth most the brightness variation occurs very close to the galaxy core and little else is seen. Some packages provide the option of writing the results to a simple text file, which means that results can be manipulated and displayed using software such as Microsoft's *Excel*.

Isophotes

We are all familiar with contour lines on maps – lines connecting points of equal height above sea-level. Isophotes are lines on an image that connect parts of the image that were of the same brightness. Often you will be able to specify the number of isophotes shown on the image or the brightness values they correspond to. For most imaging they are of no real assistance in visualization, but when applied to comets with tails or jets, an isophote plot can highlight delicate detail.

Polar Representation

Normally, when an image is displayed it is shown in a rectangular (Cartesian) representation, that is to say with the normal x (horizontal) and y (vertical) axes. So if we, for example, display a globular cluster we get the normal ball of stars. But

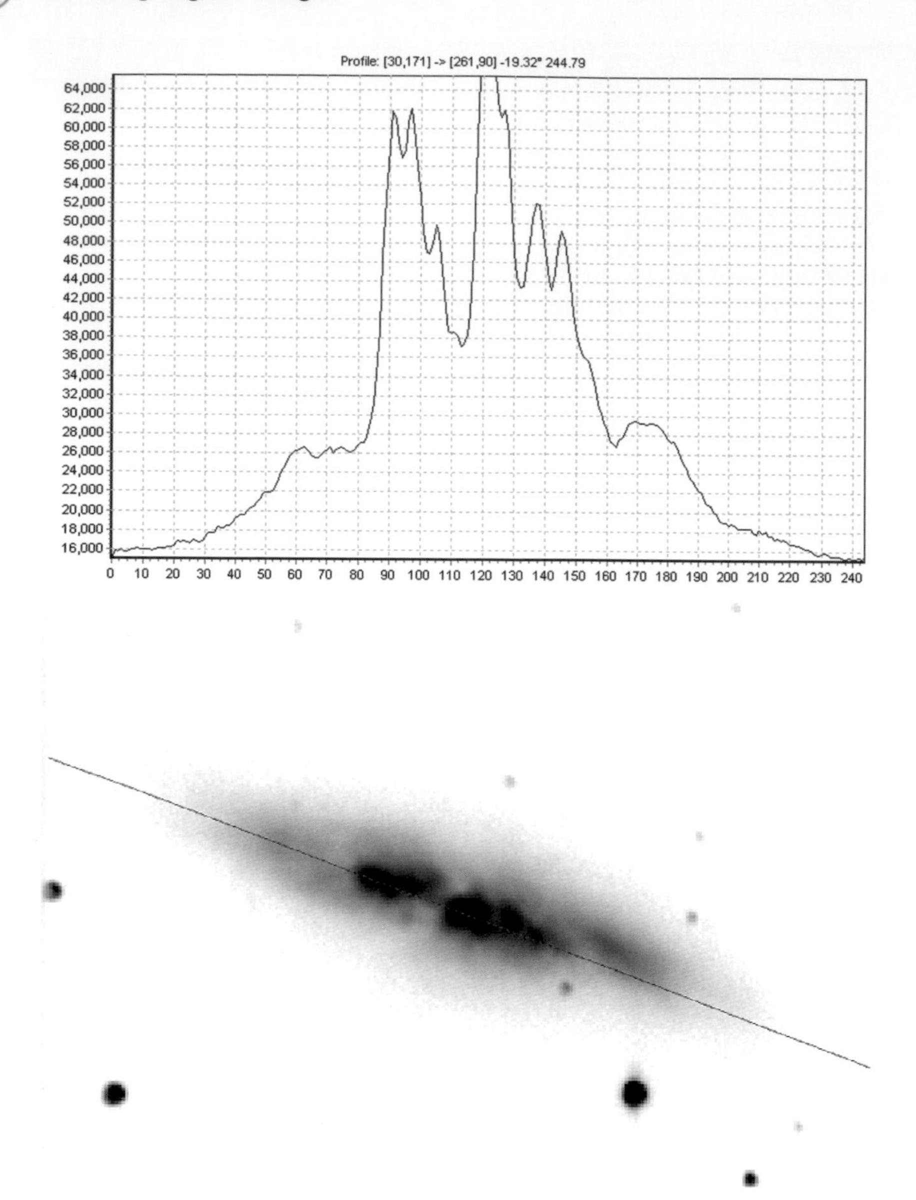

Figure 8.4. The "Profile" function found in many packages allows you to create a graph showing how brightness varies along a line drawn on the image. In this instance a section through an M82 image is shown. Image credit: Grant Privett.

there is another possible representation. If we mark the center of the globular with the mouse, and then select the "Rectangular to polar" option (or something similarly titled), the image is transformed into a polar representation. In this case the *y*-axis becomes the distance from the center of the cluster and the *x*-axis becomes the angle of a line drawn from the globular center to any particular pixel, the angle

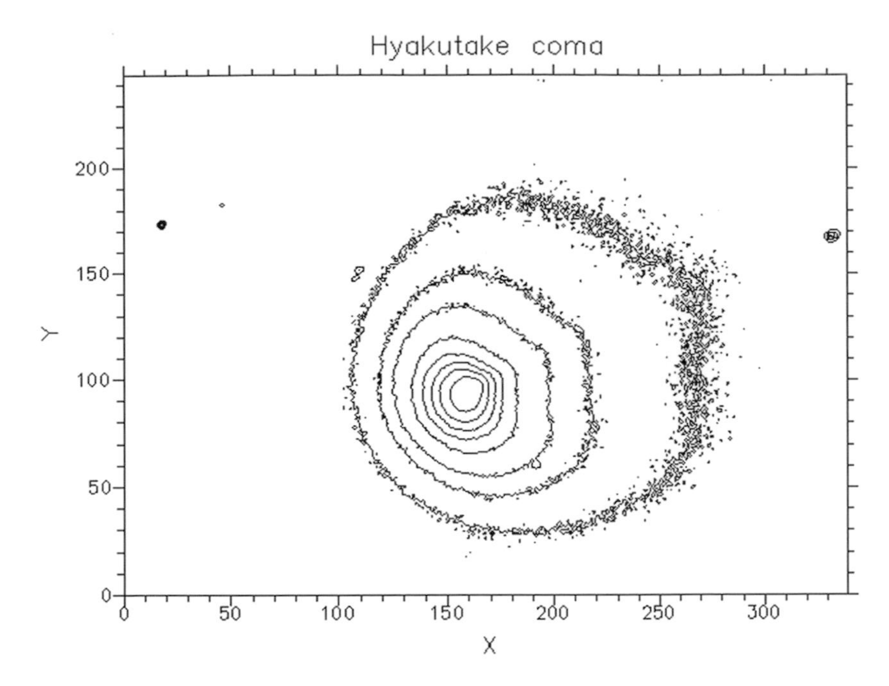

Figure 8.5. The extraction of isophote contours within the coma of comet Hyakutake. The selection of thresholds for the contours has a large impact upon the final result. These contours were manually selected and not evenly spaced in intensity. Image credit: Grant Privett.

being defined with reference to the top of the original image, or north. So if the globular is circular (most are slightly elliptical), you would get a bright area running evenly along the length of the *x*-axis with the brightness fading as you move up the *y*-axis and – in this representation – further from the center of the cluster.

It takes some thought but Fig. 8.6 probably helps to make it clearer what is going on. The potential value of this capability is not immediately obvious, but used against a face-on spiral galaxy or planetary nebula it can help bring subtle ellipticity to the fore, while if used against a planet like Neptune you can locate close-in moons more easily. If a false-color palette is employed it can create an image that is nearly an abstract art form. There is normally an option to go from polar coordinates to rectangular. I cannot say that I have ever used it, but others may find it useful.

Bas Relief and 3D

Many people like to visualize a 2D image as a surface where the brightness of a pixel defines the pixel height above the surface. So for a flat background image you have a flat surface, small galaxies form broad raised humps and bright stars give an effect of tall spires rising from the surface. This can be quite interesting to look at, particularly for a feature-rich image such as a globular cluster or for an image of the Moon, where the results can be striking. Some packages now give the opportunity to not only display data as a 3D surface viewed at an angle, but will also

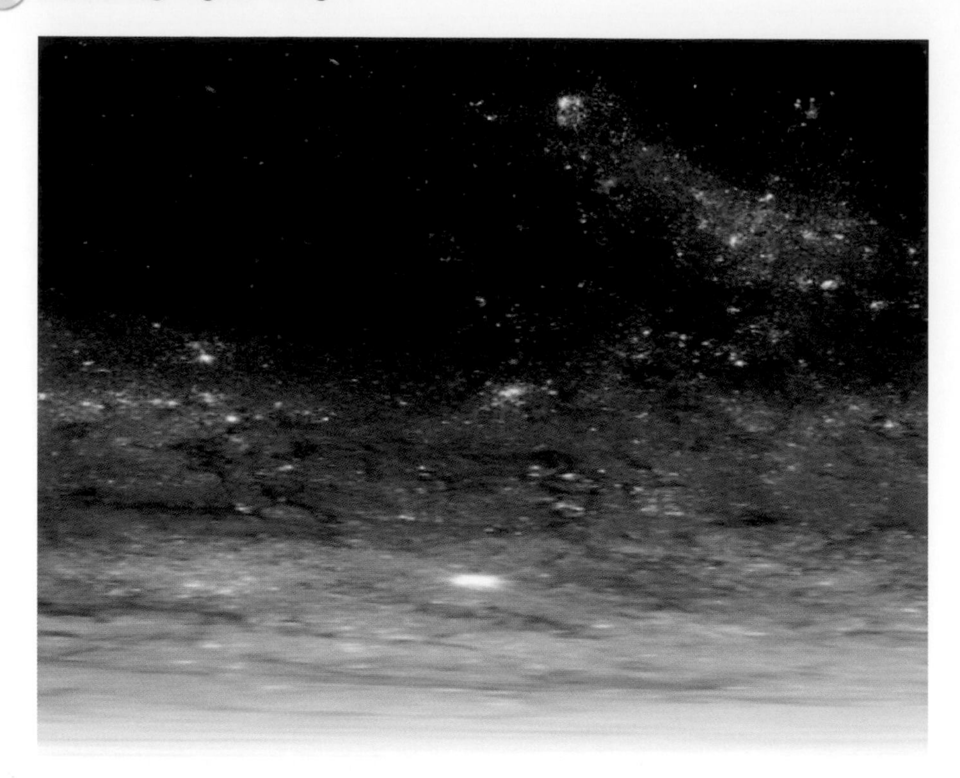

Figure 8.6. A section of the result produced when you convert an image of the glorious spiral galaxy M101 from rectangular to polar coordinates. The bottom of the image is the galaxy core with left-right representing the angle outward from the core and the up–down axis representing distance from the galaxy core. The bright blob is a foreground star. The image shows roughly a quarter of the whole. Image credit: Original image NASA+ESA.

rotate it in real time as you move the mouse or depress a button. Very impressive stuff.

The "Bas relief" operator is also available within some packages. It is quite an uncommon expression these days and refers to carved surfaces – and that's pretty much how your image can look after you apply this operator. Going back to the visualization of an image as a 3D surface, the operator modifies the image so that the highest (brightest) parts cast short shadows upon the neighboring parts of the image. The image is then displayed as if the surface was from directly above. The effect is a little odd, but can be effective at revealing quite low contrast parts of the image. Its use is not as popular as it was a few years ago.

Blinking

Blinking is the practice of displaying two images of the same scene one after the other in a single window. If the two images are perfectly aligned and of the same exposure, the vast bulk of the image will not change. What will change will

Figure 8.7. These images of M81 were processed identically apart from the application of DDP at a late stage. Note the slight flatness of the image, but also the improved clarity of the detail within the inner regions. Image credit: Peter Shah.

be the noise. Both images will have noise, but being a random process no two frames will be the same. At this point you will undoubtedly be shouting "Well, what's the point then, Privett?" but what it gives you is a perfect system for identifying the presence of novae, supernovae, variable stars, asteroids and Kuiper Belt Objects (KBOs). Any of these will stand out like a sore thumb. A nova or supernova will flash on and off as you blink the two images, highlighting its presence. An asteroid will appear as a star that moves distinctly from image to image. A KBO is more likely to appear as a star that shifts very slightly from side to side.

Obviously, the technique only works well on images that are well aligned and most packages insist on this being fully accomplished before the images are selected. But others allow you to fine tune the image alignment, providing dialogue boxes that you click on to move one of the images up–down or left–right. Blinking works extremely well and has been used by astronomers trying to discover new asteroids and supernovae. A number of packages also allow you the option of blinking more than two images so you instead see a short sequence – a sort of time lapse image sequence. This can be especially useful for looking at very faint objects that may or may not be noise. The virtue of this approach becomes apparent as soon as you start deep imaging anywhere near the Milky Way. Searching there for novae manually really is like looking for a needle in a hay-stack, with hay-stack searching being a preferable option, sometimes.

DDP

Digital Development Process (DDP) is an image enhancement process that attempts to make images appear more as they would have looked had the imager used film emulsion. The software modifies the brightness of pixels to make it easier to see detail within the brightest parts of the image (the central portion of a galaxy, for example) and to obscure the noise present in the dimmest parts. It then emphasizes the detail to bring out subtle features that might otherwise have passed unnoticed.

Many people find that this process generates very pleasing results – certainly in the image shown in Fig. 8.7, more can be seen. However, some others feel that the images created can be a little lacking in dynamics. Ultimately, like so much of image enhancement, it is a matter of opinion whether the image has been improved. The answer is most simply summed up as "the one you like best".

Edge Detectors

These seem to be supplied as options with nearly every image processing package. They highlight areas of an image in which the brightness is changing quickly, and are implemented in a variety of ways – some of the best-known of which are Sobel, Canny, gradient and Prewitt filters. Some even distinguish between vertical and horizontal features in an image.

The overall result can be very interesting to look at, yet despite this their practical use is pretty limited, unless you are undertaking some form of self-brew unsharp mask exercise. To make matters worse, the results are frequently less than spectacular until a logarithmic or histogram equalization scaling is applied to the resultant image.

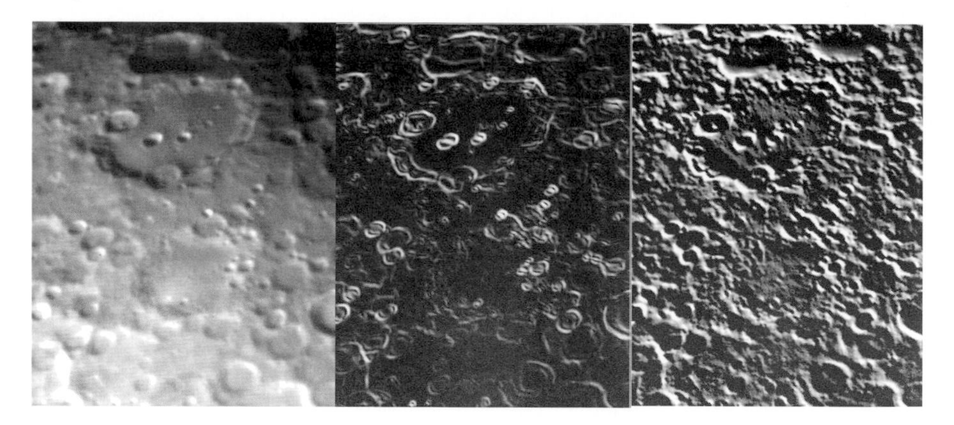

Figure 8.8. The effect of two simple kernel filters upon an image of Clavius. The image on the left is the average of five images taken with a 50 mm (2 inch) achromatic refractor in poor seeing conditions. The image on the left is unfiltered. The image in the middle is the result of a Sobel edge detector. The image on the right is the output of a gradient filter.
Image credit: Grant Privett.

Figure 8.9. The observatory and sky photographs combined to make this image were taken using a standard SLR camera and film and subsequently scanned. They were processed separately, and then combined taking particular care to ensure stars were not visible through the clouds. The observatory photograph was taken during the day. Image credit: Nik Szymanek.

Masks and Blending

Those of you familiar with normal digital cameras may be familiar with using programs like *Photoshop* to create image montages; perhaps showing several people at a party, or particularly pretty or memorable scenes from a holiday. These can be created as overlapping pictures or, alternatively, blended together so that the boundaries of the pictures are less abrupt and jarring to the eye.

This idea can be extended to the combination of astronomical images. Creating simple montages as for everyday photography is easily done using a simple cut and paste operation. But a slightly more sophisticated approach is showing an astronomical image in a human context. Examples would include people, small telescopes or observatories silhouetted against a starry background and can be found in many astronomy magazines and even *Burnham's Celestial Handbook* from the 1960s, where the analogous effects were created using darkroom techniques.

To combine images well, you will need to employ the "Mask" and "Blend" options within your image processing software. These differ quite a lot from program to program, but they will all share the concept of "layers". A layer is an image. A mask is another image – usually grayscale – which is associated with a given layer. The mask defines which parts of the layer will be used when images are combined. A number of layers and, hence, masks, may be used to create the final image – the number depending upon the required effect. The way in which the images are combined is controlled by the information in the masks and also the blending technique selected by the user. Creating the mask can be quite an effort, but the rewards can also be substantial.

CHAPTER NINE

Image Enhancement

We have reached the point where the laborious image reduction stage is behind us and we have an image created from our mass of darks, flats and images. This is the point at which to backup the image both to hard disk and also to some removable media such as a DVD, USB memory stick or CD, as a permanent record. You will find that adopting a consistent approach to filenames that involves the date and object name will save a lot of time later. In the old days of *Windows* 3.1 or DOS the naming issue was worth thinking about, as a name could be at best 11 characters long. No such limitation now exists, so give all the information. Something like:

Date-OBJECT-scope-camera-exposures.FIT

or similar, contains much of the information you might need, in the title itself. An example might be:

20060103-ngc2903-lx200-sbig8-10x60s.fit

where the image is taken on the January 3rd, 2006 – I used a YYYYMMDD format to allow directory/folder contents to be displayed chronologically. The object targeted was the galaxy NGC 2903 in Leo. I used a Meade LX200 and SBIG8 CCD camera. The exposures were of 60s long and there were 10 of them. As you can see the format is quite compact.

The FITS header can then supply the rest via its comment and data fields.

We are now undeniably at the point where the real fun of image processing begins and many options become possible. This excitement is tempered by the knowledge that progress can become slow – working out the best combination to select from the enhancements available is not always obvious. Do not be surprised if, once you are experienced, your image reduction phase takes but a few minutes and the rest of the evening is given over to image enhancement.

When working on your images be aware that nearly all image enhancement techniques render an image useless for photometric purposes, and, at best, decidedly iffy for astrometry, i.e. scientific purposes. We will be discussing astrometry and photometry later. However, if you are careful to backup every image, you will always have an unenhanced image to go back to, should you wish to revisit them later.

On a general note, during image enhancement you will probably find that the action you undertake most is hitting the "Undo" button. To make life easier, most software packages will have a hot key sequence such as CTRL-Z or F9, which means that if you press CTRL and Z simultaneously, or the F function key, the last action you carried out on the image will be reversed. In some packages you may be able to go back several processing stages, so it's a very powerful tool.

Background Gradient Removal

A background gradient – a gradual brightening of the image background as you move from one edge of the image across its middle to the other – is an image artifact that can arise from a number of sources. Common causes include the presence of the Moon in the sky, illumination of the sky by towns, illumination of the telescope from nearby lights or even haze. If you are imaging something very bright, such as a planet or the Moon, a subtle gradient across the image will have little impact upon your final result, as it will be very faint compared to your target. By contrast, if you are, instead, imaging a large faint diffuse nebula or a whole constellation, even a slight gradient can become quite obvious, especially when you adjust the brightness scaling to bring out the faint real detail in the image. If the gradient is strong, it can make it impossible to create a good quality picture without one edge appearing black and the other edge white.

One sure way to highlight the presence of such a problem is to apply a histogram equalization. If that shows the background sky as pretty much the same brightness throughout the image, then no gradient is present.

Fortunately, several programs such as *AIP$_4$WIN 2.0* provide simple ways to remove, or minimize, the gradient. They attempt this by creating a model of how the background gradient varies across the image and then subtracting that from the original image. Ideally, the image background count should then become zero, on average. If you choose to display the model image created, you will find that it will look rather like a sky image with no stars on it.

The important thing is how the software decides which bits of the image contain only sky, and which bits are stars or larger non-sky features such as nebulosity. A few programs are totally automated in their approach, but spectacular failures are quite common, so the manually assisted approach remains popular – especially for the software developer, who will find it considerably easier to implement.

Some programs allow you to select small regions of the image that you believe have no vestiges of diffuse nebulosity or very bright stars. These regions are then examined and a number of statistical criteria used to estimate what the background count really is in that part of the image, with the resulting numbers being fed into the model creation.

Still other programs allow you to manually mask out regions of the image that you believe contain nebulosity or scattered light of some form and so must not

Figure 9.1. An image of the field containing the very small and dim Princes Nebula. The top copy shows the impact of light pollution upon the image background. The lower image shows the same scene after the removal of a simple bilinear plane. Image credit: Grant Privett.

contribute to the calculation of sky background values. It then samples the rest of the image as it wishes, to create an accurate model.

As a general rule, where the Moon is very distant from the field of view or the telescope is pointing to near the zenith, it will frequently be the case that the gradient is linear, i.e. the background brightness variation is smooth and neither accelerating nor slowing as you move across the image. Such a simple variation is often modeled as a 2D "plane" (as variation may be occurring at different rates in the x and y directions) and subtraction of the model sky gradient from the original is usually very successful.

The situation can be more complicated than this, as more than one source of light generating a gradient may be present. For example you might be imaging an object low above the horizon with a glow from a town off to one side. One of these may generate a linear variation but the other might lead to a brightening on one side of the image that increases more rapidly as the image edge is approached. In such a case a "polynomial" model may be required. This approach leads to a model that may fit the data very well, but is normally sensitive to where abouts on the image the small background sampling locations are placed upon the image. Because of this, it is normal to place the points in a simple evenly spaced grid, using as many points as you can. In some cases a successful result can sometimes be obtained by placing the sampling points around your target object in concentric circles.

Even worse, you might have a telescope which is illuminated intermittently by lights from the house next door, and be imaging with the Moon above the horizon. This can lead to very complex forms of gradient which take a correspondingly more sophisticated model and lots of background estimations to eliminate. The problem then is that the fit will seldom be perfect and may leave some low-level artifacts or gradients of its own. You will encounter times when you have a background variation that you just cannot remove. This alone is reason to ensure your telescope is never exposed to direct light from anywhere but the sky.

Several general rules apply to selecting the locations on an image at which the gradient modeling software will sample the image. These can be summarized:

- there should be few stars present;
- the points should not be within nebulosity or galaxies;
- avoid the edge of the image where residual vignetting may be present;
- keep away from bright stars – scattered light will be present;
- try to make them evenly distributed – a grid is good;

Figure 9.2. The unsuccessful removal of sky brightness gradients. The images show the before and after result of removing a simple bilinear plane from a sky gradient that was, in part, generated by a nearby light source, rather than distant towns. Image credit: Grant Privett.

- use as many as possible;
- avoid anywhere that flat-fielding doughnut artifacts can be detected – though you should have got rid of those through flat-fielding!

If the above rules are employed you will generally increase your chances of a successful gradient removal. As to the type of model employed, try to keep to the KISS rule: "Keep It Simple, Stupid". In general, start with the simplest linear model and only move on to the more complex forms when the simple forms fail.

It is worth noting that gradient removal software is often available as a "plug-in" – software run from within a particular program, but not written by the authors of said package. A well-known example is *Photoshop* which has a considerable number of such add-ons, including a very useful FITS format reader and at least one for gradient removal. It is worth looking on the World Wide Web to see if one is available for your chosen software. The software authors will often collect up available plug-ins or provide links to them from their website. The downside is that the "plug-in" software is not guaranteed to work, but if you have ever looked at the purchase agreement for *Windows* XP you would be amazed anyone ever buys software, for virtually nothing is guaranteed. Use them with caution but expect to find some very nice free, or nearly freeware.

Image Sharpening

These are generally simple "filters" that are applied to an image to increase the ability of the user to see detail within it. Often you will find them described vaguely as "Sharpening", "More Sharpening" or more specifically as "Laplacian", "Hi-pass" or "Difference of Gaussian". Individual packages differ in how these methods have been implemented – there is no one option that is suited to all type of imagery – and the names merely serve to emphasize the fact they will each have a different effect upon an image.

If one type of sharpening is too strong and another too weak, it may be worthwhile creating a new sharpened rendition of the original image and "adding" it, or some proportion of it, to the original image using the "Mathematics", "Add" or "Numeric" functions of your software. These serve to let you effortlessly combine aligned representations of your images to the best effect. So, if you wanted to add 10% of a sharpened image to your original, you would sharpen a copy of the original, multiply it by 0.10 and then add that to the original image. This can yield very pleasing results, although you may find unsharp masking (of which more later – it is a form of sharpening despite the name) has a very similar effect.

In some software packages you will find options providing "adaptive" or "FFT" filtering. Depending on the types of images considered these can lead to fewer artifacts and so are well worth investigating. The way they are implemented by the software author can vary, so you should never assume that, say, the filtering techniques in *AstroArt* are the same as those used in, for example, *Photoshop*. Be prepared to (yet again) experiment a lot.

For those keen to understand Fourier transforms and able to cope with mathematics beyond that found in school lessons, some references are provided to processing in the "frequency domain". For most people the concepts involved are

Figure 9.3. A demonstration of the effect of low- and high-pass filters upon the final image. The image on the left is the average of five images taken with a 50 mm achromatic refractor in poor seeing conditions. The middle image has had low-pass filtering applied, thereby blurring the image. On the right-hand image high-pass filtering has been applied, revealing much detail. Image credit: Grant Privett.

rather stretching, and are certainly not trivial, but getting to grips with the ideas can be very rewarding – you will never look at an image or sine wave the same way again.

One particular occasion when an FFT comes to the fore is when there is a regular repeating pattern of interference within an image. In such a case convert the image to its FFT image equivalent, look for any especially bright spots within the FFT image and blur or reduce their brightness in some way, and then convert back to an image from the FFT. This can work well, because a regular pattern on a normal image will be represented by sharp bright point(s) in the FFT domain. However, modifying FFTs can be very sensitive work, so be ready to use the "Undo" button a lot.

Image Smoothing

It comes as a surprise to many astronomers that you would ever want to blur an image – surely that's the sort of stuff that made us so lovingly pay attention to the focus of each image we take. The most obvious reason for doing so is a poor image scale. Imagine you have created an image where each pixel is greater in width or height than the size of the current atmospheric "seeing". In such a case, the telescope focal and pixel size length might be such that a pixel is 4 arc second across and all but the brightest stars would appear as single pixels – albeit bright ones. Such situations can easily occur with short focal length instruments and cameras system where the sensor has big pixels. To give the star images less of a "pixellated" look you can apply a slight "Gaussian", "Soften" or "Blur" filter and these will spread the light from each pixel slightly to give the image a less jarring look.

Care should be taken in choosing the blurring option selected as some utilize a simple square shaped filter, which will result in stars of a "boxy" appearance. This

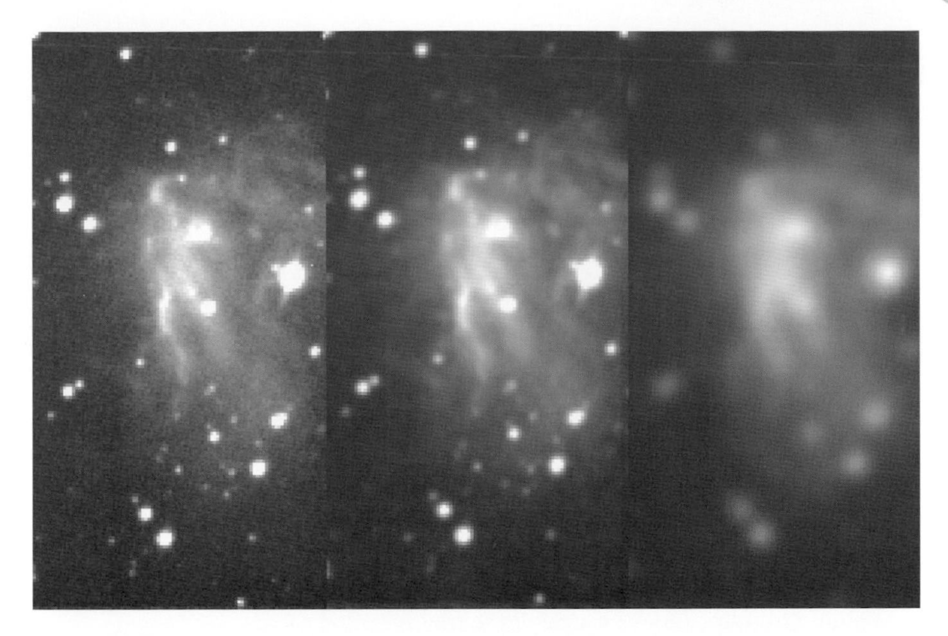

Figure 9.4. Examples of image smoothing. Images are, left to right; the raw image, Gaussian smoothing over a one pixel radius and over a 10 pixel aperture. Image credit: Grant Privett.

can be avoided by employing a "Gaussian" or "Soften" option where the size of the filter is expressed as a radius rather than a width.

In some cases you may find it works best to simply resample an image, i.e. increase the image size in terms of pixel numbers by a factor of 3 before blurring it and then resample it by a factor of 0.333 to the original size. Often you will find that the resampling operation has some user defined options, and among these may be bicubic spline, cubic spline, bilinear interpolation or others that are similar sounding. These options can improve the appearance considerably, with a bicubic spline being a popular choice.

Median Filtering

Earlier, during the image reduction section, when describing image stacking we mentioned a technique called median stacking. We noted that it was often applied where lots of possible values had been measured for a given pixel and the aim was to select the value most likely to accurately represent the pixels' true values.

A conceptually similar, but distinct, process is used to reduce the impact of noise on a *single* image. In median filtering a not dissimilar approach is taken to determining if any of the pixels of an image are spurious, by examining those immediately surrounding it. Spurious values can arise from cosmic rays impacting the sensor, from poor processing during the image reduction stages or where the exposure was too short. They are often applied when the user wants to make a random background "quieter", to permit a subtle feature that is nearly submerged in image noise to be seen more clearly.

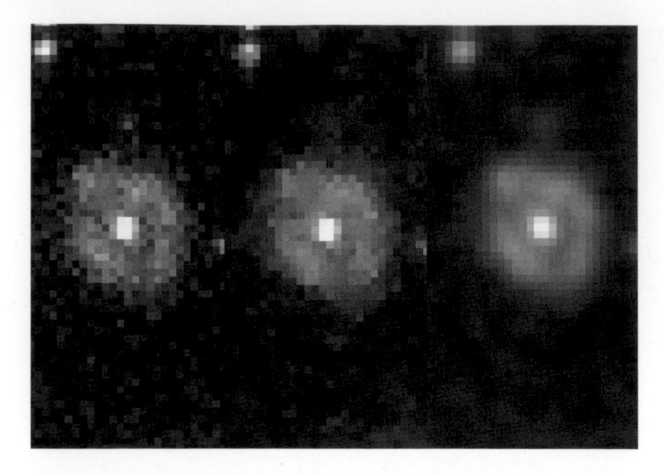

Figure 9.5. The result of median filtering and averaging upon a noisy image of a planetary nebula. The images show, left to right; the raw image, median filtered or averaged/mean. Image credit: Grant Privett.

Median filtering operates roughly as follows:

1. look at a pixel and obtain its value A;
2. look at all the pixels surrounding it;
3. deduce their most likely value, B;
4. compare A to B;
5. if A compared to B is brighter by an amount exceeding a threshold value, put value B into pixel A;
6. repeat for every pixel in the image.

At first, this sounds complicated, but what it amounts to is asking the question "Is this pixel absurdly bright compared to its neighbors?" and, if so, changing its value to something more representative of that part of the image. With an image that has poor signal-to-noise, this will work wonders to make detail more obvious, and will frequently remove smaller cosmic ray hits as well. The algorithm can be applied using lots of different threshold value settings, with recourse to the "Undo" key, until you feel the noise has been suppressed sufficiently and the detail you wanted to see has been made more apparent. Be warned though, the price paid for this action is a tendency to blobbiness – not a technical term clearly, but descriptive – within the scene. Use the filtering too strongly and large diffuse features may be revealed, but at an unacceptable cost in lost faint detail.

The parameters you are likely to be offered, prior to median filtering, will be a width or radius and a strength/sensitivity. The first is pretty much universal. Options are generally 3, 5, 7, 9 or *N* pixels. As an example, if a value of 5 is used then $5 \times 5 - 1$ (24) pixels will examined to deduce what value the middle pixel could be replaced by. *N* allows the user to apply even bigger filters. When the value becomes large, most of the detail within an image will be lost and the effect becomes rather like applying a big Gaussian blurring. The other parameter, if requested, refers to the sensitivity of the routine. It, in effect, defines how much brighter than its surroundings a pixel must be, before it is deemed necessary to replace it. Set it to too low a threshold and too many will be replaced, generating a blobby textured background – not really what we want. On the other hand, set it to too high a threshold and nothing much will happen. The human eye is surprisingly sensitive to the blobbiness, so it is best to err on the side of caution.

It's a constant source of confusion to those new to astronomical image processing, but no amount of median *filtering* of a single image will create the same effect as median *stacking* multiple images of the same scene. Median stacking works by extracting data from lots of images and so has more information to play with. Median filtering has the information in but one image to utilize and so can never do as good a job – if all the exposure times are the same.

Unsharp Masking

Unsharp masking is a term that arises from the time when all astro-photography involved dark rooms, noxious chemicals and painstaking hours working on a single frame of emulsion to squeeze every iota of data from it. In essence, the digital implementation of the technique is one that can best be described as highlighting detail within an image, but only if the scale of the detail is smaller than a certain user-defined size. An example might be the Orion Nebula, M42, the outer regions of which are dim, wispy and very beautiful. Unfortunately, the inner portion is also hugely interesting, with lots of detail, with the bright trapezium grouping of stars thrown in. To show detail in both the inner and outer portions, a technique is required that makes detail in both regions visible, despite the very different brightness levels. Unsharp masking is such a process.

It works like this: you select two (sometimes three) parameters, which go by different names in different packages – examples are sigma, coefficient, radius, clipping, etc. One of them relates to how strongly the detail within the original image will be emphasized, while the other relates to the scale of the detail you can sharpen. The third parameter, if present, will define a threshold, which limits the enhancement only to parts of the image where it has a significant impact. Once these are supplied, the software creates a copy of the image that has been deliberately blurred – strange but true. This will lack some of the fine detail in the original but will retain the big broad features. It then subtracts this blurred image – or some adjusted version of it – from your original image. This means that, in the output image, any features that are smaller than the blurring size will be enhanced substantially, while larger scale variations will be suppressed.

Now, it may not be immediately obvious that anything involving image blurring is a useful thing to do, but looking at the results quickly dispels any doubts. Certainly, no planetary or lunar imager would want to be without it and those seeking detail in all manner of deep sky objects would be seriously hamstrung without its ability to bring out detail in the center of globular clusters, the arms of galaxies and even planetary nebulas. As with all image enhancement, you can overdo things and this quickly becomes obvious when the image starts to become grainy. Images with high signal-noise ratios do best – the Moon is a prime example – but even then the process can be overdone, so that the images stop looking sharp and appealing, and degenerate into a mess of stark harsh edges unlike anything you see through the eyepiece.

Figure 9.6. The effect of unsharp masks of differing strengths upon the final image. The image on the left is the average of five images taken with a 50 mm (2 inch) achromatic refractor in poor seeing conditions. The middle image has had unsharp masking applied. On the right-hand image, far too much unsharp masking has been applied.
Image credit: Grant Privett.

The best way to judge when to stop is by carefully examining the image to see when it stops looking natural. You will know you have gone too far when dark rings start to appear around bright stars. These rings will be of a scale similar to the size of the blurring mask employed on the original image and can be very apparent, especially if there is a background sky gradient.

If you have any doubt about your last unsharp mask operation, then you probably went too far. As with other image enhancement techniques, you can get away with more over-processing on large images because, when printed out, the faults are rarely as obvious as when scrutinized on a screen.

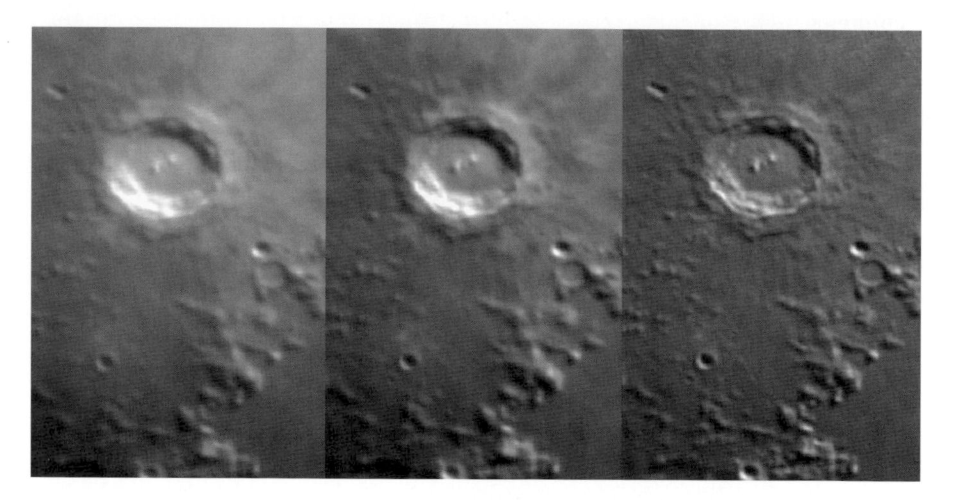

Figure 9.7. The effect of unsharp masking on a slightly blurry image of the Moon. The left-hand image is untreated. A 50 pixel wide unsharp mask was applied to the middle image. The right-hand image had 50 and 5 pixel unsharp masks applied consecutively. The intensity scaling is linear between the lightest and darkest pixels. Image credit: Grant Privett.

Generally, unsharp masking even a megapixel image takes at most a few seconds to accomplish, so the technique can be easily applied as a quick look-see when first processing an image, and looking for fine detail.

Image Deconvolution

This title covers a range of sophisticated image processing techniques that are employed to sharpen an image and make it more pleasing to the eye.

The deconvolution techniques you will find most popularly employed within image processing packages will undoubtedly include Maximum Entropy, Lucy–Richardson, Weiner filtering and Van Cittert. The mathematical basis of these individual techniques is fascinating – accounts can be found in the book *Digital Image Processing*, by Gonzalez and Woods, detailed in "Further Reading" – but the workings of the algorithms are well beyond the scope of this book. Suffice to say, they use a computationally intensive approach to try to estimate how the image would have looked had it been taken using a perfect telescope located above the Earth's atmosphere. To do this, the software normally looks at the profile of a user-selected star on the image and considers that to be representative of how the telescope optics and atmospheric conditions have modified and blurred the original image. This is true because a star is so far away that, from space, it should appear as essentially a single bright pixel. But, instead, it appears as a small bright circle surrounded by a dim diffraction ring, or even an animated, writhing, amorphous blob. With this information, the software then redistributes some of the light within the image in an attempt to reconstruct the undistorted appearance of the light sources – it is usually an iterative approach.

Those of us who recall some of the first high-quality images returned from the *Hubble Space Telescope* (HST) will remember how successfully deconvolution techniques such as Maximum Entropy – the trendiest technique of its time – were

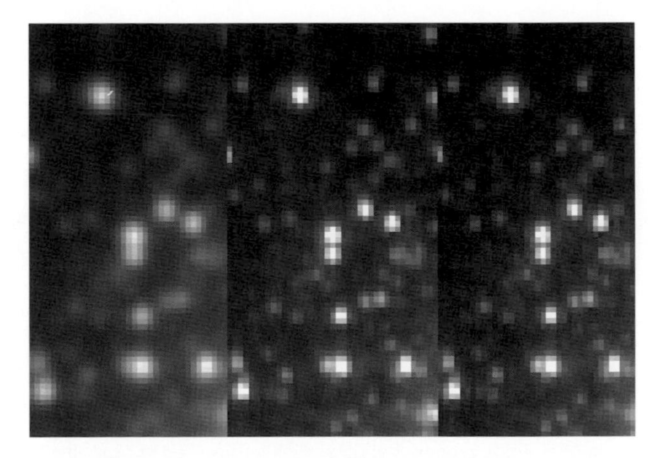

Figure 9.8. The application of two deconvolution techniques to a high SNR image of the outer regions of the globular cluster M15. The techniques applied were, left to right; the raw image, Weiner filtering and Van-Cittert. Both the algorithms do well. Image credit: Grant Privett.

employed on images of Saturn. These helped to overcome the blurring caused by the telescopes misshapen main mirror. At the time, the operators of the HST gave little publicity to the fact that, for images with lower signal-to-noise ratios (as is often the case with leading edge research), deconvolution techniques have problems. They can easily fail spectacularly, turning random noise in the background into objects, creating artifacts and generally making a pig's ear of an image as an accurate account of what was seen. For a few years, the only research projects the HST could work on were those involving bright targets, and many research projects and researchers were left high and dry as a consequence.

Another reason to be careful with deconvolution techniques and to use them only in moderation is that, for reasons we will not go into here, there is again a strong tendency for stars to develop dark halos around them. They can also amplify background noise, making the image look grainy. These effects are commonly most evident when the deconvolution process is allowed several iterations of processing, as this will be pretty certain to yield visually unappealing results strewn with artifacts. Some deconvolution techniques work by making an attempt at processing the image (first iteration), and then making another attempt using the result from the previous attempt (second iteration). This can be continued to the third, fourth or twentieth iteration, with each subsequent iteration using the product of the previous iteration as its input. Fortunately, you do not have to set this up, you merely specify the number of iterations required.

Personally, I cannot remember ever using more than 12 iterations of any deconvolution technique, but that can depend upon how the software implements the basic algorithms involved, and how the method has been adjusted to allow more

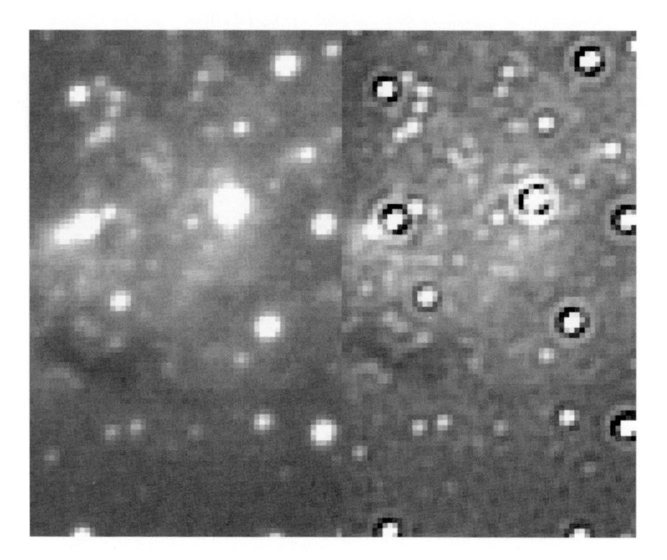

Figure 9.9. The application of the Maximum Entropy and other deconvolution algorithms has a tendency to create dark halos around the stars in an image. Left to right: the raw image and the Maximum Entropy processed image. The dark halos are most apparent around brighter stars lying amidst a light background. With careful adjustment of the parameters employed, the number of iterations executed can help minimize them. Image credit: Grant Privett.

subtle application between iterations. Look out for software parameters that soften or mute the change from iteration to iteration. These can help a lot.

Used delicately, these techniques can help to bring out clumping in galaxy arms or highlight the belts of Jupiter. The artifact problem is slightly less apparent if the image lacks an illuminated background, as the dark rings then appear against an already dark background. Large images also help to hide the unsightly rings, as they become more difficult to spot.

It is tempting to treat deconvolution with extreme caution, but as always, the best bet is to experiment and see what you find yields the most appealing images. Snap judgements as to what works best should be avoided and it can be useful to keep several variations of an image showing subtly different detail to hand, and choose between them some days after they were created. Certainly globular clusters, spiral galaxies and planetary nebulas all benefit from this approach.

CHAPTER TEN

Handling Color Images

There is a problem with imaging and image processing. It is addictive. A nice picture almost immediately rewards the effort taken, by reminding you of the beauty the universe has to offer and the fun you can have seeing it – I have one above my desk at work. And the only thing better than a good image is a good color image. Be warned, once you start taking and processing color images you may not want to go back to monochrome again.

Acquiring Color Composites

As was mentioned earlier, the quickest route to creating color images is to employ a single-shot color imager. These are very attractive and are made more so, since the newer generation of cameras boast large and sensitive sensors providing very wide fields of view. Another advantage, from the image processing viewpoint, is that, in effect, you can take one exposure, but get all the color information needed in one go. In places, where most nights are clear nights, this has no impact, but under the cloud strewn British skies there is no way of being sure that a clear night will last. So, if you planned 20 minutes worth of red images followed by 20 minutes of green and then a final 20 minutes of red, you would be completely stymied if after 30 minutes it went cloudy. You would not have enough green images and you would have no blue images at all. That would mean no final color image. However, with a one-shot camera, every frame contains the essential information. You might not have the SNR you hoped for and the image might be a bit grainy, but there would be something solid to show for the 30 minutes of effort.

Unfortunately, one-shot cameras do not produce the very best quality color images. There are two reasons for this. The first is that, because of the filters incorporated within them, the pixels can never be as sensitive as those in an unfiltered monochrome camera. The second is that the final color assigned to a given pixel will be an estimate based upon the intensity of light that reached both it *and* some of the surrounding pixels.

The best images are generally created using red, green and blue filters placed in turn in front of a monochrome CCD; that is, the combination of three images taken through R, G and B filters to form a composite color RGB image. An RGB image is formed by using the relative intensities of the R, G and B images to provide the color information for each of the pixels, while the sum of the intensity at each pixel is used to define its brightness in the output image.

To be entirely fair I should say that the difference between filtered and one-shot color imaging is becoming less marked than it was, but the option of unfiltered imaging gives a flexibility that one-shot users may sometimes envy.

This "true color" approach is rivaled in popularity by an alternative approach which creates so-called LRGB composites using four images – the standard RGB filtered images, plus one unfiltered luminance (L) image.

The use of four images combines the advantage of the greater sensitivity of the unfiltered image with the color information contained within a set of RGB images. They are quite fun to make, as most software now allows you to effortlessly combine images from different sources. For example, I have combined a luminance image I took using my telescope, with the RGB images taken by a friend using an entirely different telescope and camera combination. The final result was very much better than the sum of the parts.

One point worth remembering is that if generating true color images is the intent, the luminance image should really be filtered to remove the infrared light which

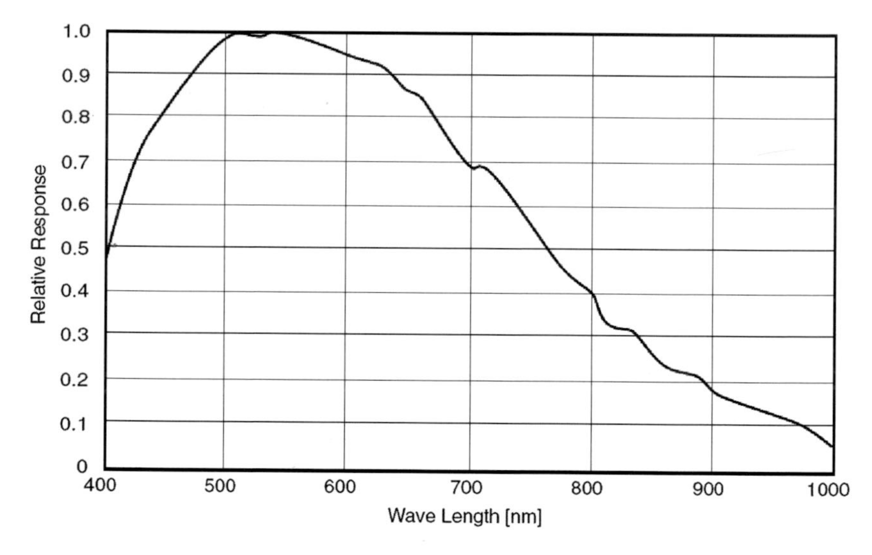

Figure 10.1. The relative response of a CCD in terms of the wavelength (color) of the light it receives. This chip (Sony ICX285) is most sensitive to light of 530 nm wavelength, with the sensitivity falling by a factor of two by the time you reach both 400 nm (blue) and 760 nm (deep red). Note the sensitivity continues out toward a wavelength of 1000 nm.
Image credit: Starlight Xpress.

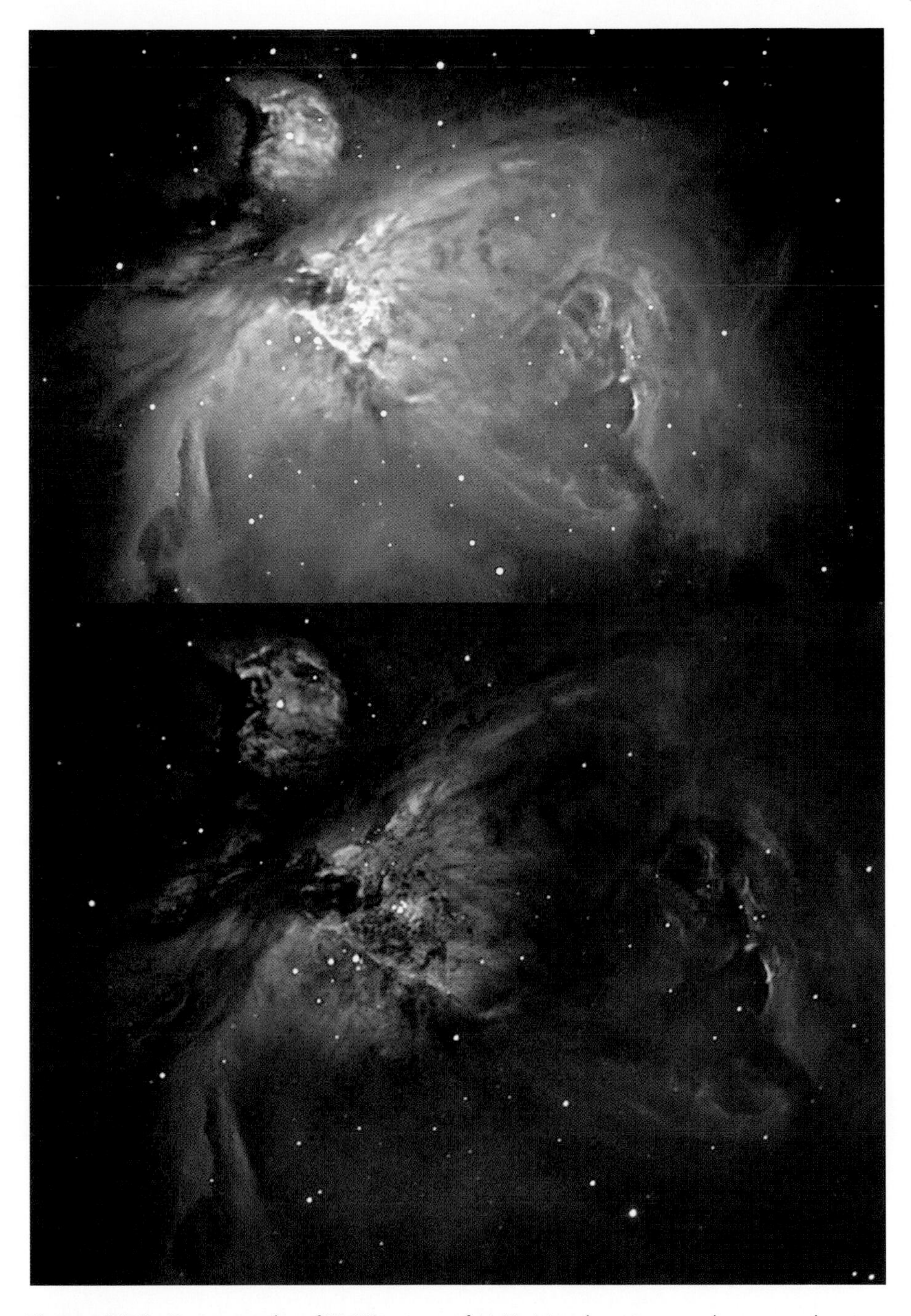

Figure 10.2. Two examples of LRGB images of M42. Note how image enhancement has been applied slightly differently, leading to quite different – but equally valid – results. Image credit: Nik Szymanek and Peter Shah.

would not be detected using RGB filters. In many cases, in particular for galaxies, the inclusion of near-infrared light makes only a subtle difference to the final result but, in some cases, the effect can be striking – a particular example is M27, the Dumbbell Nebula, which is transformed when the infrared light is included. Indeed, taking images through filters that pass only infrared – despite their disconcertingly black appearance – can be a rewarding pursuit in its own right.

Processing Color Composites

Ignoring the obvious difficulties involved in getting the right color balance with RGB or LRGB images (which will be discussed shortly), processing monochrome images is rather easier than color RGB or LRGB images, for a number of reasons. The first thing to remember when taking color images using filters is that, by using a filter, you have made it necessary to have a flat-field for each of the filter colors you are employing. If this is not the case, the dust doughnuts created by dust on the filters will not be properly compensated for in your image reduction. As might be expected, a flat-field taken on one night might not be right for another night, as some fresh dust may have arrived and/or old particles fallen off. This means extra complication, but the effort is greatly minimized by the automated processing options mentioned earlier.

The most obvious of the other difficulties encountered when working in color occurs when one of the images you capture contains an asteroid, a cosmic ray, meteor, aircraft or satellite. The presence of any transitory object will generate a strongly colored artifact if appearing on the R, G or B frames or possibly a whiteish artifact if present on the L image. Such an artifact will be very conspicuous and may ruin an image. One way to avoid them – and they happen frequently – is to create each of the individual component L, R, G or B images from several images, using image stacking that employs the median or sigma stacking methods. This way, if one of the frames being stacked is corrupted, then the stacking method should readily create an output image without the artifact. It does have the side-effect that you may end up with an image of slightly lower SNR, but the alternative is to try to edit out the artifact on the original image, which can be very time consuming and is unlikely ever to be very satisfactory.

One of the other issues that may be encountered when taking images of the same target with different filters, is that of focus. We all know that, as the temperature changes at night, the length of a telescope changes slightly and thus refocusing may be necessary. But it is less obvious that the individual L, R, G or B filters may bring light to a focus at slightly different points, even at a stable temperature. This creates stars surrounded by small colored annuli and is due to the use of filters made from glass of slightly different refractive indices. This is especially true of filters that are not purchased as part of a comprehensive set from a single manufacturer.

The simple way of overcoming the focus shift is by carefully refocusing after every filter change and after every 30 minutes imaging. The problem can also be addressed by ensuring that the filters purchased all have the same "optical depth" and so focus at the same point – in the same way that you can buy sets of eyepieces that require no refocusing when swapped.

If all this effort fails, or you are combining images taken on nights with different seeing, you will indeed find that bright star images are surrounded by small, sometimes subtle, colored rings. This should not be confused with the blue fringing found when using achromatic refractors and with some types of camera lenses. One way to address this is to enlarge or resize the image by a factor of 3 and then attempt to gently sharpen the most blurred image, or blur the sharpest image, so that the artifact is minimized. The resultant color composite image can then be resampled to the original resolution if necessary. Obviously, degrading any image with a blur goes against the grain, but it can improve the overall appearance in such cases.

The final problem that may need to be addressed is that of noise in the individual filter images. As the use of any filter must inevitably reduce the signal-to-noise ratio of an image, the resultant RGB image may be a bit grainy. In LRGB cases where this occurs, taking an unfiltered luminance frame using full sensor resolution and creating the individual RGB images in a binned 2×2 pixel mode will improve the SNR. The lower resolution RGB is then resampled/sized to the same size as the luminance image and all the images are combined. Perhaps surprisingly, the difference in initial resolution rarely becomes obvious in the final LRGB image and many striking images have been produced in this way.

Due to the way they are made, sensors are more sensitive to some colors of light than to others. In the 1990s, the sensors most commonly found in amateur cameras had very good red sensitivity, quite good green sensitivity and poor blue sensitivity. Because of this, it was quite common to use longer exposures when imaging through a blue filter, compared to the other two. So you might collect light for 1 minute in red, 1.2 minutes in green and 2.8 minutes in blue. An alternative approach was to create your G and B images by stacking and co-adding more B and G images than you used for R.

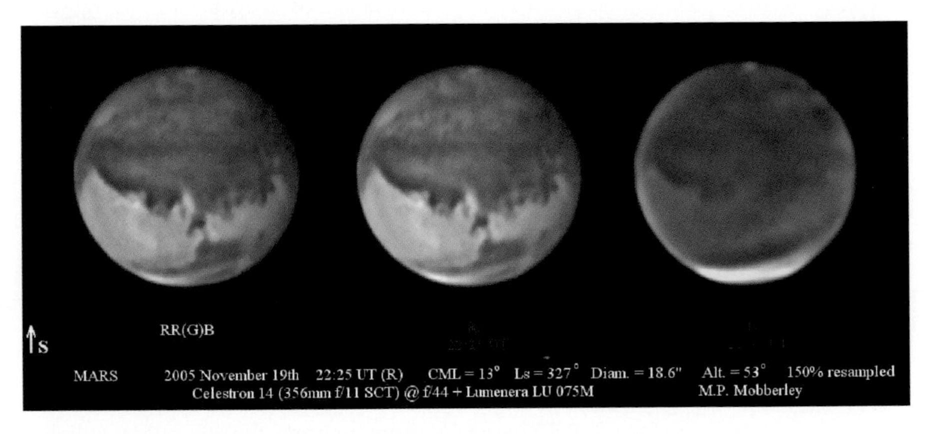

Figure 10.3. An image of Mars from the 2005 opposition, this time using a 356 mm (14 inch) SCT operating at f/44. Interestingly, in this composite, the red image has been employed as the luminance image and a "green" image has been approximated by averaging the red and blue images. Consequently, instead of creating a traditional LRGB, we have an RR(G)B composite where (G) denotes the approximated image. This approach means that only red and blue images need be collected. It saves a lot of time. A similar technique can be applied to Saturn to overcome the weak blue content, but it is less successful with other planets. Image credit: Martin Mobberley.

The situation is less clear-cut than previously, as many modern cameras are better at detecting blue light than they were. But despite this, for any camera there will still be some bias. It is less important than it was, and will become even less critical as sensors improve further, but be aware of the issue when imaging with individual filters. A final technique that can be used in processing is the application of a gentle median filtering. Gentle is the important word here. Beware, blobby images that way lie.

In an earlier chapter we mentioned that composite images taken through cyan, magenta and yellow filters (hence CMY) are created in the same manner as RGB composites, i.e. by taking images through three separate filters. You will find that this works well for imaging most astronomical targets but you may find that, in some cases, it is difficult to achieve a color balance that you are happy with. That said, the results can be very successful.

Achieving Color Balance

If you have a color image of any form it is essential that you examine it with a critical eye to determine if the colors look true to life. With the planets and Moon this is not a problem, as we can look through the eyepiece and see how they really look. With that view in mind, we can then adjust the balance on the screen to ensure – say – that Saturn's globe is dusky yellow, the Moon silver-gray and Mars tinted ochre. But things are decidedly trickier when dealing with the deep sky. It could fairly be argued that with the exception of the pale blue-green appearance of some of the brightest small planetary nebulas – like NGC 7662 – no one has ever seen much in the way of color in the deep sky, so it is not immediately obvious how to adjust the color balance; in which case, we are forced to take the colors exhibited by background objects as our cue and hope for the best in getting things right.

For example, if you look at an image and find that all the stars are tinged predominantly red, green or blue, then something is probably wrong. In a wide field you can expect a mixture of yellow, red, white and a few scattered pale blues, with white or yellow the most commonly encountered colors. If the field of view the image takes up is quite large, you may even be able to identify one of the brighter stars from a star catalogue and determine its spectral type. The Sun's is of type G2 and, like other G stars, appears white-yellow, while F, A, O and B stars are increasingly bluish. F stars such as Procyon, Canopus and Polaris appear to be essentially white. Similarly, K, M, R and N stars are from types that change from yellowish to red. All these stars might then be used as a reference for setting the color balance, with yellow and white stars the best choice. Suffice to say, if an M type star appears blue in your image you know something is seriously wrong.

As red, green and blue filters do not transmit exactly the same amount of light, and because the sensitivity of your sensor will depend upon the color of light being observed, you will find that combining images of equal exposure with each filter does not give sensible results. Most stars will appear to have a color bias toward the red. Fortunately, we have two alternative approaches that allow us to get the colors right. The simplest is to use images of the same length exposure and then weight them – so that the contribution from the image in red (where most sensors are very sensitive) and, to a lesser extent, the green, will be played down. The alternative is to have equal weighting, but exposures of different length for each color.

It is possible to deduce the appropriate weightings or exposure length ratios by imaging a sheet of clean, but unwashed white linen under skylit conditions. Paper, paint and washed linen are not as good a choice, as the manufacturers of paper and soap powders add ingredients that fluoresce in sunlight (due to its ultraviolet content) giving a false result. The weighting required depends upon your sensor but, unless you are using a very good DSLR, you will have to apply some weighting to get a good balance.

It is worth bearing in mind that even if your RGB exposures are balanced carefully to ensure an accurate color rendition, and even if you are using a single-shot color camera, the image color balance may need adjustment if the field being observed is less than 30 degrees above the horizon. This arises because everything will appear to redden as it approaches the horizon – just as the Sun and Moon do when setting or rising.

Remember, also, that if you image from a heavily light-polluted site, ensuring that the image background appears black may be difficult. Use of the "Curves" function with each of the individual RGB frames to adjust their weightings at light levels near the background is the simplest solution – but can be fiddly.

Some digital cameras have a number of lighting condition modes. Canon use the term AWB (automatic white balance) which gives six options for different types of illumination and also allows "custom" modes that can be set up using a test card. These can be very useful.

Another serious issue requiring processing can be the color cast of the sky background. When imaging from light-polluted skies, the background sky can take on a distinct hue due to a sky glow generated by a mixture of mercury or sodium lights. This will look pretty unattractive and makes achieving a visually pleasing final image difficult.

Fortunately, unless the object of interest is of a brightness very similar to that of the sky background – as can happen – it will usually be possible to adjust the

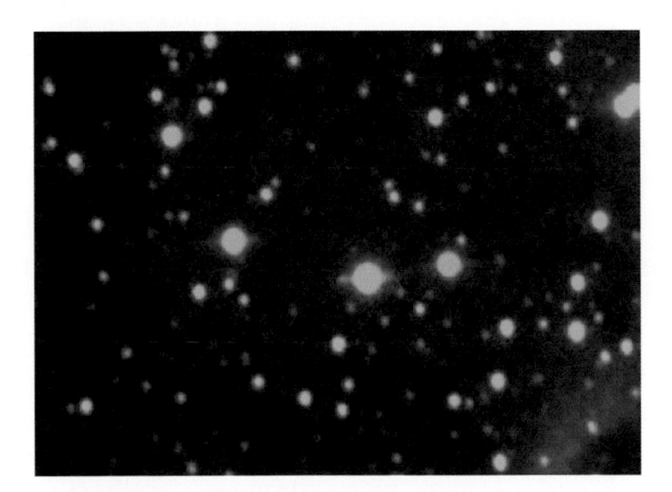

Figure 10.4. An RGB image where the exposure through the red filter was too short, so that the whole image has a color cast. Modifying the weightings of the filters would help overcome this. Image credit: Grant Privett.

scaling method employed, and the black/white thresholds, so that the overall color cast of the background is not obvious. The "Curves" function mentioned earlier is especially useful for this purpose. Often, by just adjusting the shape of the curve at low light levels the sky is rendered a uniform black or dark gray, leaving the color balance of brighter parts of the image essentially unaffected.

Narrow-Band Filter Composite Images

It is worth mentioning a form of composite imaging that allows you to see where, within an image, light of a precise color is being emitted. An example of such a color filter would be the H-alpha (or H-α) filter, which only permits light emitted by hydrogen in a particular state to reach the imaging sensor. The most popular and perhaps best-known application of narrow-band H-alpha filters is in imaging the Sun, where the filters reveal lots of detail that is totally invisible in the unfiltered image. Similarly, an H-alpha image of the Orion Nebula makes visible much detail that would otherwise go unseen. H-alpha filtered solar telescopes are rather expensive bits of kit, but hydrogen filters suited to deep sky imaging are rather cheaper to buy and are owned by a considerable number of imagers.

Imaging with a single narrow-band filter can produce dramatic monochrome images, especially with objects like supernovae remnants, emission nebulas and planetary nebulas. By indicating the location of gases of certain types, a narrow-band image conveys scientifically useful information, but the resultant color schemes are unlike those you see for simple RGB or CMY filtered images and sometimes produce quite surreal and abstract effects. This is one of the reasons – apart from being imaged by a multibillion dollar 2 m orbiting telescope – that many HST color images make such an impression; the bold brilliant coloring grabs your attention and then the subtle detail keeps you interested.

All types of chemical elements emit light in a range of colors across the electromagnetic spectrum and many of these colors can be selected for imaging using narrow-band filters. H-alpha refers to a specific red color of light emitted by hydrogen. Other filters are listed in Table 10.1.

To create narrow-band composite images you need to combine light from three narrow-band images using, for example, H-alpha as red, O-III as blue and S-II as green.

One option is to use three separate H-alpha, S-II and O-III images to create a composite image akin to a RGB image – it is probably just as well no one calls such an

Table 10.1. The most commonly used narrow-band filters

Name	Source element	Color
H-alpha (H-α)	Hydrogen	Red
H-beta (H-β)	Hydrogen	Blue-green
O-III	Oxygen	Blue-green
N-II	Nitrogen	Red
S-II	Sulphur	Deep red

Figure 10.5. An iconic picture. Part of the Eagle Nebula. A narrow-band filter image – possibly the best-known example of a composite image. Image credit: NASA/Hubble/ESA/ASU.

image an HS_2O_3 or it could be confused with chemistry. Alternatively, you could create an LRGB, using an unfiltered image as L and then employing the H-alpha, S-II and O-III images to provide the RGB content. This raises a ticklish question; which of the filtered images should be used as indicating red, or green, or blue? There is a tendency for imagers – perhaps influenced by the color schemes used by landmark HST images – to use H-alpha light as green, S-II as red and O-III as blue, but others use S-II as blue, H-alpha as green and O-III as red instead. There is no "correct" choice – apart from trying to ensure the background of each image is not affected by an obvious color cast – as even professionals vary in their treatment of narrow-band images.

Overall, narrow-band images can be handled in a manner very similar to what we use to exploit for RGB images, but with some important differences. In most astronomical color imaging there is an underlying expectation that many stars (if unsaturated) in an image will appear as white, or nearly so. But in narrow-band imaging this is not the case and so you can freely apply gamma stretches and other

Figure 10.6. Two images of the Pelican Nebula, taken using the same telescope and CCD. The left-hand image is a normal LRGB composite. The right-hand image is an LRGB where the images treated as R, G, B were obtained through hydrogen-alpha, sulphur-II and oxygen-II filters. Image credit: Peter Shah.

histogram adjustments to the individual images. This makes life very much easier as – again – there is no right or wrong answer and the final image need only be aesthetically pleasing. The alternative is to let the color balance be defined by the intensity of light received through the filters, but that would mean that in a composite where most the light came from, for example, an H-alpha image, everything would appear to be of essentially one hue. Whilst interesting, that's not an especially attractive prospect and would mask otherwise invisible features. Happily, by simply increasing the brightness of the other two images you get to see more clearly where light of other colors is being emitted.

A particular advantage of this type of imaging is that narrow-band filters very strongly filter out, and thereby suppress, the background skyglow caused by light pollution or moonlight – sodium and mercury filtered light excepted, perhaps. This dramatically improves the contrast of faint diffuse objects and means that narrow-band imaging is frequently undertaken by observers living in urban locations. It is particularly successful in urban environments because it makes some form of high-quality deep sky imaging feasible.

There is a drawback – there had to be really; anyone who has put a narrow-band filter to their eye and looked at the sky will be able to tell you that the stars fade a lot; often three magnitudes or more. Clearly, much of the light received is just discarded, which has a severe impact upon the SNR of the images obtained using them. This means that images can appear quite grainy, so the familiar median filtering or Gaussian blurring and median filtering can again be useful techniques to employ to disguise the graininess. There are two other options that are worth trying; the first is to construct a color image using an unfiltered image as luminance and the narrow-band images for color information. While this will help to hide the graininess in the narrow-band images, it somewhat negates the point of narrow-band imaging, by producing a hybrid result. On the other hand, if it produces an attractive image, what is wrong with that? The second approach is to not use a luminance image and merely bin the narrow-band images 2×2 so that the signal-to-noise ratio improves. This can make individual star images rather small and undersampled – but on a wide-angle shot this can be an acceptable price to pay.

Perhaps the simplest approach to improving the signal-to-noise ratio of your narrow-band images is to take the longest possible exposures. Certainly you will find that the exposures needed to get a good result, with a narrow-band filter in use, are of at least 10 minutes in duration. This can be achieved by stacking multiple images, but a more satisfactory – though more expensive – approach is to use an auto-guider to guide the telescope and track the stars accurately during an exposure.

Figure 10.7. Composite LRGB color images of the Pelican Nebula created using light from sulphur, oxygen and hydrogen as the R, G, B color components. The L component was an unfiltered exposure. The images show the results of combining the filters in different orders. Starting top left and going clockwise the orders are: RGB, BGR, GRB, BRG.
Image credit: Grant Privett.

CHAPTER ELEVEN

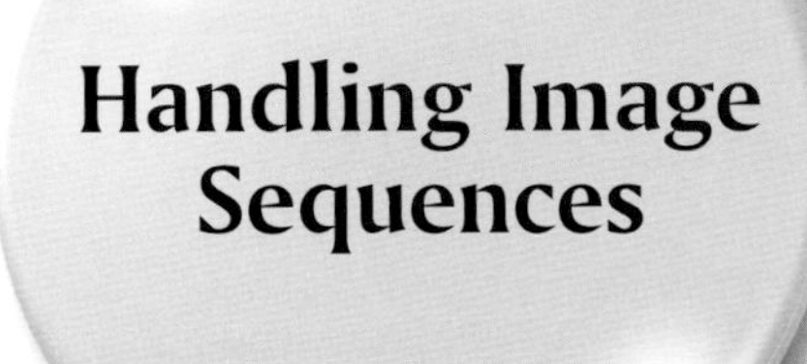

Handling Image Sequences

Image Acquisition

A lot of what follows relates only to the use of webcams for acquiring images, but the principles apply equally to images from some DSLRs and increasingly from images from other sources, such as the Mintron camera.

In general, acquiring image sequences is reliably handled by the software that is supplied with your webcam or CCD camera. But in recent years the authors of some image processing packages (such as *Maxim-DL*) have begun to provide the option of camera control. This can even include the use of webcams that have been modified to allow long exposures, but be sure to read the software specification details *very* carefully before buying your software. Mistakes can be expensive.

The most important thing you then need to do is to ensure that the images are created in the manner and format you would like to deal with. Because using webcams for astronomy is very much a fringe pursuit, the default settings for software supplied with a webcam are rarely appropriate to doing astronomy. For example; you may find that the images created have been reduced in resolution due to pixel binning, have been JPEG compressed or have been sharpened before you get your hands on them. Whilst the defaults chosen by most webcam manufacturers will be great for showing Aunty Enid's 70th birthday party to family via the web, in astronomy every photon is precious and artifacts of any sort to be avoided like the plague. For astronomy you should configure your camera – whatever it is – to full resolution, no filtering, no compression and the maximum color range/depth. It is even worth ensuring that any filtering the software employs to cope with low light level noise (frequently a simple median filter variant, such as the salt-and-pepper filter) is very much turned off. As with normal astronomical imaging, you will need to ensure that you take dark-frame images so that you can apply them later.

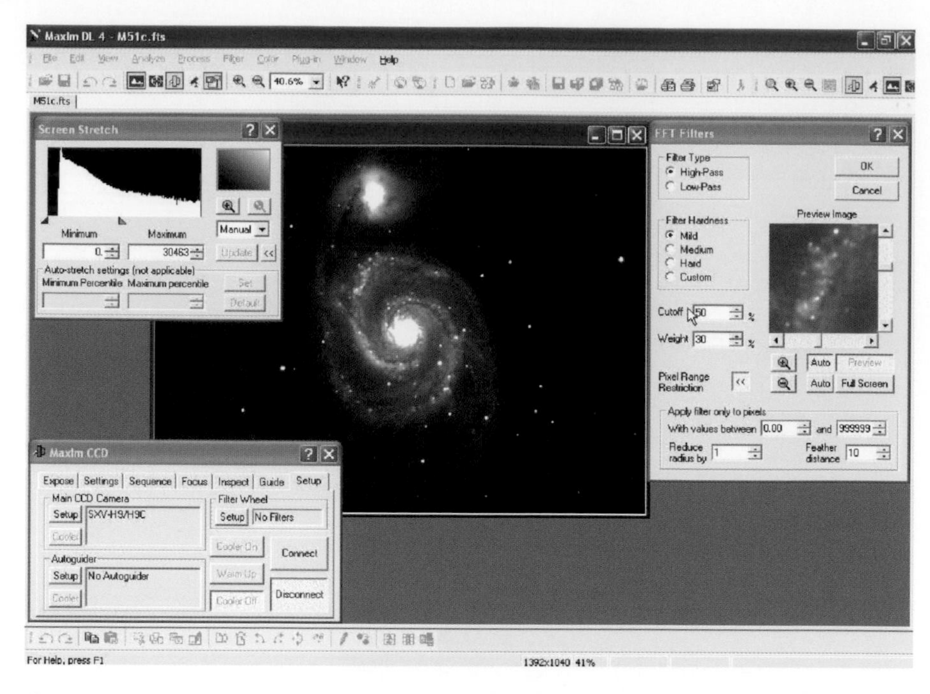

Figure 11.1. The very powerful *Maxim-DL* package. An impressive, versatile and well-designed, but quite expensive, piece of software. Image credit: Ron Arbour.

Those of you using USB 2 enabled cameras of any type will find it easier to get pictures that have no compression artifacts, but the rest of us still using USB 1.1, or even parallel port driven software, have to be more careful. The problem is that if you set a high frame rate (more than five frames a rate) you may find that the images will have been subjected to a lossy (some detail permanently lost) compression technique, applied when the image is transferred from the webcam to the PC. For this reason astronomical imagers using USB 1.1 cameras tend to use a slower frame rate, so that all images created are transferred unfiltered. Obviously, USB 2 has an advantage in that its ability to transmit lots more unfiltered images per second make it the current interface of choice – no doubt there will be a USB 3 or Superwhizzo-USB along in due course. A very high frame rate does, however, have the disadvantage that the camera will gobble up your disk space at an alarming rate. In general it's worth trying to ensure you have a fast (7200 rpm) disk with more than 80 Gb of free space, so that a good night is never brought to a premature end. Users of DSLRS in particular may want to buy large memory chips for their cameras, to avoid the need to swap them out at regular intervals as the nights' work mounts up – an external power supply for a DSLR is also especially attractive.

It should be remembered that some current software has trouble ingesting MPEG or AVI imagery files larger than 2 Gb in size, so be sure to check this before creating very large imagery files. Large image sequences can often be edited by normal video-processing software – such as Pinnacle's *Studio* – to manually

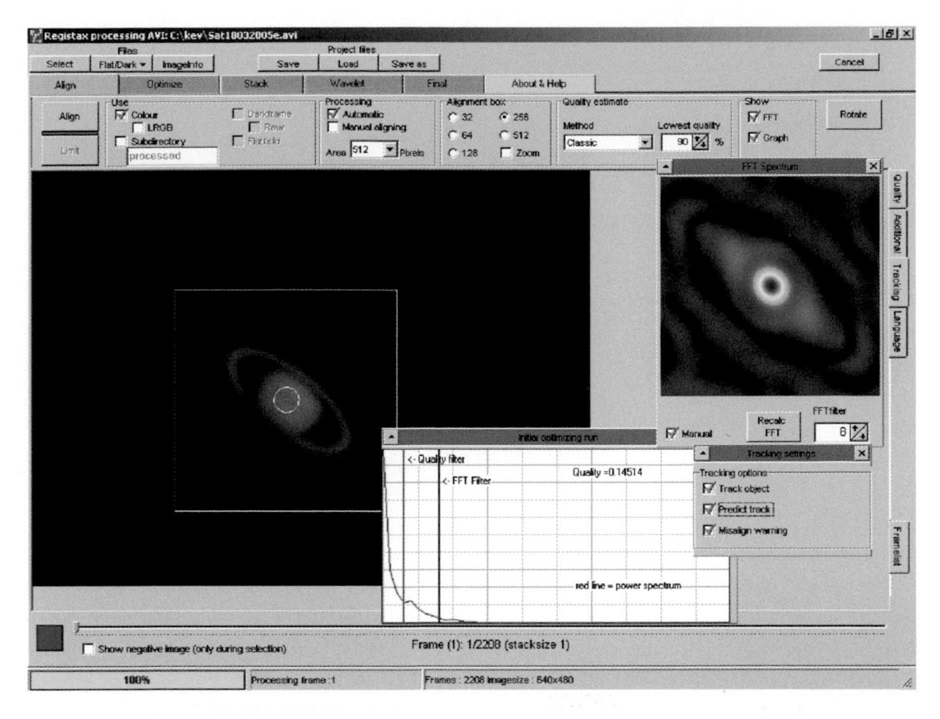

Figure 11.2. *RegiStax*. In effect, the industry standard for creating planetary images from webcam images. It is inexpensive, versatile and flexible, but can be a bit bewildering for new users. Be sure to read the documentation carefully to get the most out of it.
Image credit: Grant Privett.

remove bad frames, but the removal of blurred or distorted images is a very laborious process that can take several iterations – especially where the sequence has more than 1,000 images. For this reason it is often best to employ software such as *RegiStax* or *AstroVideo* to select what they believe to be the "best" images from the sequence.

Image Quality Techniques

In recent years the software used to select good images has become increasingly sophisticated and so rarely throws out a good image. However, success rates differ from package to package, with the major difference being in the number of poor images that are retained. To automatically determine which images are best, the authors of the most successful software packages have made a variety of methods available. Unfortunately, the names of the menu options that select them will almost certainly be far from self-explanatory. Yes, it's experimentation time again, but remember this doesn't all have to be done in one go. Run all of the available options on the same data set and look at the AVIs created, to see which has retained the lowest proportion of poor images.

In practice, it is best to be ready to spare a lot of time to work on one good set of imagery and try to familiarize yourself with the options. Examining and testing a single sequence of 1000 good images can take all evening (a fast PC helps) and is not to be rushed.

As mentioned, the software packages approach the problem in slightly different ways, employing differing levels of ingenuity and achieving differing degrees of success. So when you are happy with one, create an output AVI containing the best 10% of images and then edit that by hand to remove the dross. There is no denying that the manual culling of duff images is a really laborious and dull process, but it is very much worth the effort if the good frames are sharp.

Just throwing away the 90% of less sharp images may seem a terrible waste, but by allowing them to contribute to the stacking, you will diminish the final product. It is difficult to imagine any sequence lasting more than a few seconds where all the images are worth keeping.

Once you have found a quality selection method that does a good job on your data, stick with it – for that planet at least. One nice aspect of the software currently available is that the best-known packages for reducing webcam imagery are quite inexpensive, compared to software like *Photoshop* or *Maxim-DL*. Some of it is even free. Another nice touch is that there are, in some cases, time-limited downloadable versions available, which allow you to try out the software before you buy it.

You will quickly find that there is a wide range of webcam processing software available which mean that if you wish to you can start out by using a relatively simple program and later progress on to those boasting more functionality and/or a more complicated GUI. Frankly, the interface of the current version of *Registax* is pretty daunting – I think my reaction was "There's an awful lot to learn", or something similar, but more heartfelt – even to those accustomed to dealing with this sort of software. However, in its favor, it does have very helpful downloadable instructions, so the more ambitious can just jump straight in at the deep end if they wish.

As you progress you will find that there are a variety of tweaks you can apply to improve the selection accuracy – depending upon the software used. A reference image is normally required at some point to give the software a strong hint of what a good image should look like when assessing the quality of each frame in the AVI. Frequently, software defaults will decree that the first image in a sequence is used as a reference while in others you will have to manually select one you like the look of. An alternative is to use your best 10% of images to create an image and then select that as a reference image. As most the distorting effects of turbulence should be absent, the new reference should work more consistently.

Determining the quality of an image has been approached by different developers in different ways, including: examining the power spectrum of each image; monitoring the image brightness; determining the sharpness of a feature such as a planet's limb and also an analysis using something called wavelets and the wavelet transform. This latter method has some similarities with the more familiar Fourier-transform-based power spectrum approach. They both attempt to discriminate the real information content of each image, i.e. how much within the image is real detail and how much of the apparent detail originates from noise or distortion.

Figure 11.3. Four images of Saturn. Starting from top left and moving clockwise they are: a typical slightly blurry image from the sequence; a sharper image from the same sequence that was used as the reference; the aligned and stacked image created by combining the other images in the sequence; the final image after the application of some wavelet processing to reveal detail. Image credit: Kev Wildgoose.

Once you have created an image sequence you are happy with, you can then carry out the other essential processes.

The most important part of the processing stage, after selecting images of good quality, is the alignment stage, where many (perhaps hundreds) of images each contribute to a stacked output. There is not a lot you can do to improve how that works (some programs allow you to choose a method of determining alignment) and a lot depends upon it. So if it provides the opportunity to frame the planet using a box (which will be the part of the image used to align all the images) be sure to make sure the whole planet is encompassed within the frame.

One other parameter you can tweak, before alignment is undertaken, is any option for the creation of an output image bigger than the original. This may involve the drizzle algorithm, which combines simple resampling of your image with an attempt to improve the resolution of an image. This may be possible because you have lots of images and the software know very accurately how well-aligned they are. Drizzle attempts to estimate the brightness that pixels would have had if a sensor with small pixels had been used – similar to super-resolution techniques found in professional image processing. This option is quite sensitive to the SNR of the input images, the degree of turbulence present, the drive accuracy and the number of images you have. So there's no promise that it will always help. But it's worth a try when working with good data.

A major concern is the overall color of the image. You will need to adjust the relative strengths of the RGB colors until things look realistic. Use the view through an eyepiece as your guide. Its not uncommon to find beginners creating images of Saturn and Jupiter that are a deep yellow, which is a shame, as the overall tint will distract the viewer from any detail captured.

Another important option that has begun to appear is a tool for removing color fringing in a planetary image. This cast can arise for a number of reasons, one of which is the proximity of the planet to the horizon when observed. It was a big problem in the northern hemisphere during the 2003 Mars apparition, when Mars had a southern declination and was observed through a lot of atmosphere. Removing this fringing is done by firstly determining which edge of the image is which color, and then interactively tweaking controls that adjust the color cast across the image. The tweaking – which amounts to slight realignment of the RGB images relative to each other – is continued until the rings/planets' limb appear as they really are through the eyepiece. That should be sharp and false color free. The movement will probably be small – a pixel or two – and the movement may be in two directions. There is a tendency to overdo things, so take great care with this. It is unlikely that all of a cast can be fully removed, so be realistic.

In many ways, the method you choose to use when processing your images is not the most important thing. More important, perhaps, is the imagery it has to work on. If you have created an image sequence on a bad night when the seeing makes the planet look like it is being viewed through the bottom of a jam jar, no amount of fiddling around with wavelet transforms or effects sliders is going to create a great image from them. All you will get is the best of a bad bunch. The options that Cor Berrevoet *et al.* have provided you with in their software really come to the fore when they have some half decent imagery to work with, so be

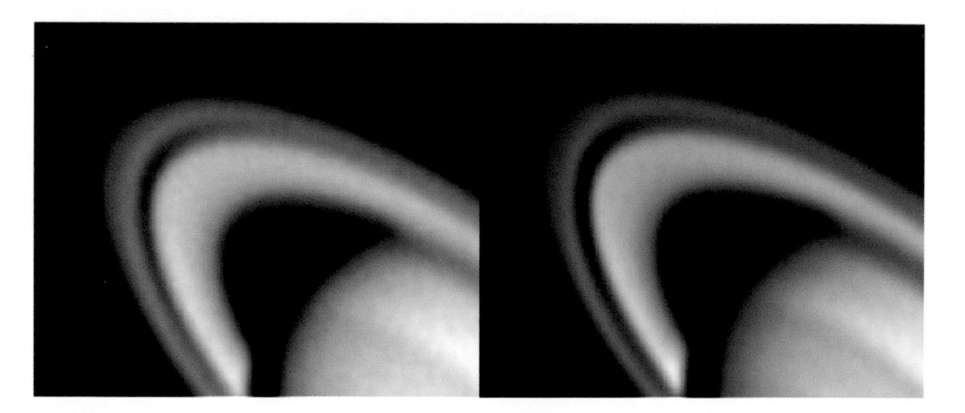

Figure 11.4. These two images show the rings of Saturn with color cast/fringing, on the left-hand image, and after correction, on the right-hand one. Image credit: Kev Wildgoose.

ready to work hard to get a decent result. There is no substitute for imagery collected by an observer who pays attention to collimation, focus and all the other factors mentioned earlier in Chap. 4, "Acquiring Images".

Packages like *Registax* are worthy of books in their own right and, inevitably, trying to give more than a feel for their capabilities is unrealistic. Fortunately, the default software settings for packages like *Registax* are selected to provide a fairly reliable output, but once you have created an image with those, see if you can improve on it. I often receive two images from friends, one is a good first draft that has got them excited and makes them want to share the news, and then, a week or so later, an improved and refined final draft.

Using modern software, webcams and normal commercial telescopes, it is possible to create images better than those ever taken by conventional photography – even using a 4 m telescope. That's an incredible thing to say, but it's true, and is just one facet of how PCs have revolutionized astronomy for amateurs. Some of the software can take a little while to learn but so does riding a bike – and how many people regret that?

CHAPTER TWELVE

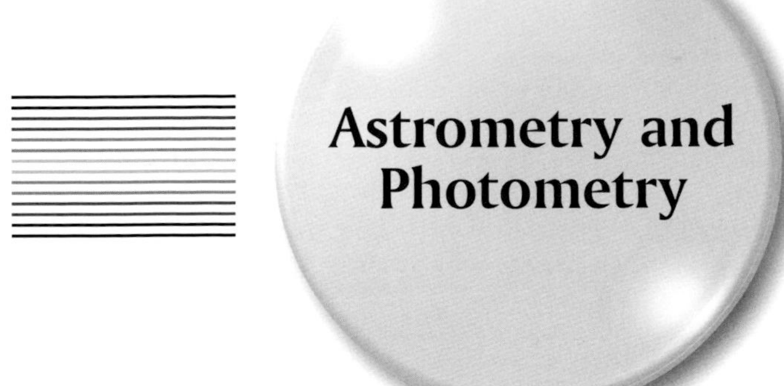

Astrometry and Photometry

Strictly speaking, these are specialist pursuits; the precise measurement of positions and the precise measurement of brightness, respectively, do not come within the scope of this book. Despite the fact they are but a part of the loose collection of techniques glorying in the increasingly inaccurate name photogrammetry – image mensuration or measurement – it would be unreasonable to ignore them and their great value.

Astrometry is generally undertaken by people who are interested in monitoring/determining the orbits of asteroids, near-Earth objects and comets, or identifying the position of newly found and sometimes ephemeral objects such as novae. Astrometry used to be done using expensive measuring microscopes to record the position of star images upon a photographic emulsion – a long and extremely laborious manual process that was worth avoiding if at all possible, if only because it was followed by an equally long and tedious calculation process. Today things are rather easier than they were, with most the measurement and all of the essential calculations undertaken by software.

Using modern software you need only read the image in, identify the coordinates of the center of the field of view, specify the size of each image pixel and the software does the rest. The output it generates often provides feedback which includes things like a precise measurement of the focal length of your telescope. Most importantly, an assessment is made of the accuracy associated with the measured position.

The bottom line is that the software notes the position of stars within the image provided, compares those positions to the positions of the stars described in the Tycho and Hubble Guide Star Catalogues and then calculates the position of everything that can be seen on the input image. This sounds complicated and indeed some of it is, but all the calculation and error estimation is done by the software, almost unaided. Depending upon how sophisticated the software is, you may be asked to look at the image and a map of the same region of sky (created from catalogues) and identify a given star on each. Not exactly a Herculean effort.

Some astrometry and image processing software now support the use of World Coordinate System data, if it is present in the FITS header file attached to an image. This makes the process even quicker. It does this by providing a record of the image scale, rough orientation (if known) and center screen coordinates. Undoubtedly, this will become the norm as makers of GOTO telescopes move into the CCD market providing integrated systems.

Several astrometry programs exist, of which the most famous must be *Astrometrica*, but some astronomical image processing software suites have some of the required functionality built in.

What you get is the ability to identify very precisely (often, to within a few tenths of a pixel size) the location of an asteroid or comet at the midpoint of the image exposure. By submitting these observations to organizations such as CBAT (Central Bureau for Astronomical Telegrams, who maintain a web-page-based list of objects, either recently discovered, or otherwise requiring urgent observations to help improve the accuracy with which their orbit or position is known) you can play a real and active part in the science of astronomy even while imaging as an amateur. It really is quite easy.

Photometry, as the name implies, is the accurate measurement of the brightness of an object. Generally, photometry is undertaken to help determine how bright an

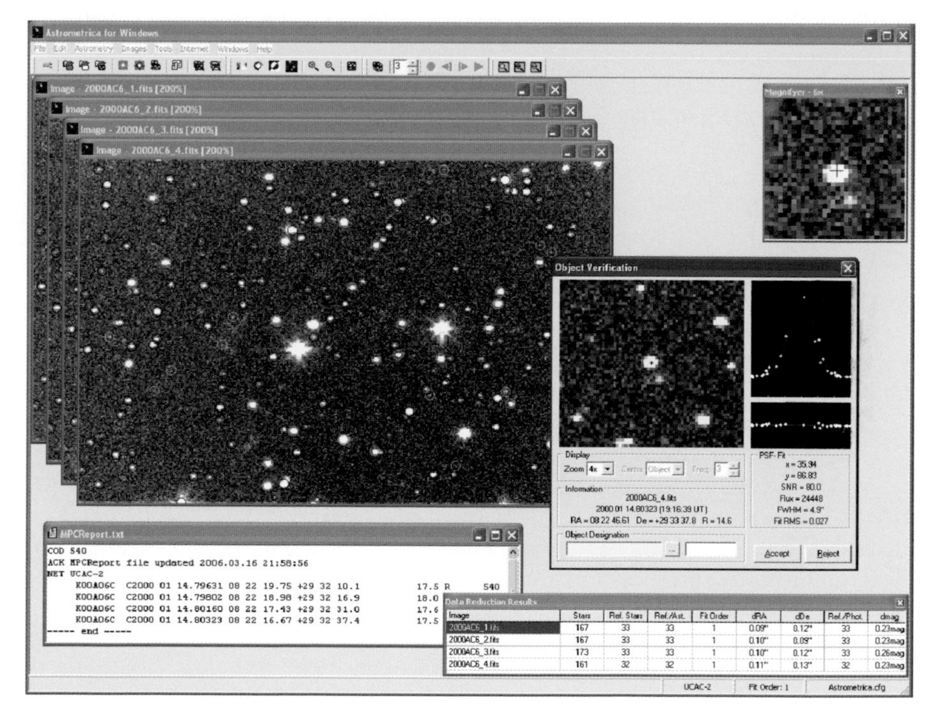

Figure 12.1. *Astrometrica* – a software suite used to provide accurate astrometric locations of stars and other objects within an image. It can be employed to extract object magnitudes by using information from star catalogues. Image credit: Herbert Raab.

object is compared to its neighbors, with a view to seeing how its luminosity varies as time passes. Particularly useful examples are the observation of variable stars such as novae, stars with accretion disks, supernovae, some variable nebulas, active galactic nuclei and even the transit of extrasolar planets in front of their parent stars – as is the case for the star HD209458. The variability in brightness may be only a few percent, below that which might be discerned by eye, but, by combining the imaging devices, software and telescopes now available, much more is possible.

Most astronomical image processing software has an option for image photometry and, in essence, they all work the same way. At heart, all the software is doing is working out how much light is contained within a star image and then subtracting from it an estimate of the light that would have been in the same pixels had the star been absent. As you might expect, the background brightness of the sky is estimated by looking at pixels close to, but not within, the star. Most software does this by drawing a circle around the center of the star image, with a radius large enough to guarantee that all the starlight has been collected. It then draws another, still larger, circle and the brightness of the pixels within the annulus formed by the two concentric circles is used to estimate the background pixel brightness. The number generated is compared with the equivalent value generated for neighboring stars of known brightness, to produce a magnitude estimate for the suspect variable.

The measured results for the variable star over several months are plotted on a graph – otherwise known as a "light curve" – showing the measured magnitude versus the date on which the image was taken. If the plot does not follow a

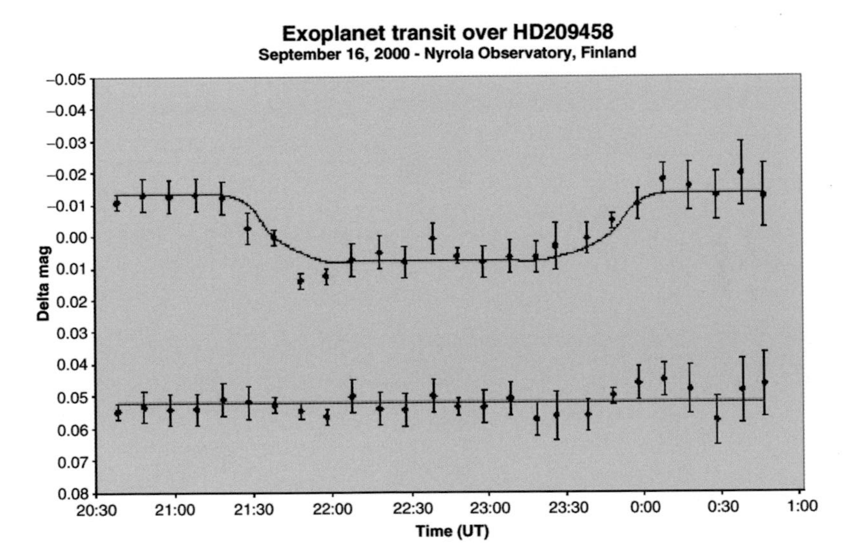

Figure 12.2. The light curve created by Arto Oksanen. The star imaged was HD209458 and the dip in the light curve was caused by an extrasolar planet orbiting it. The total drop in light is quite small, but was clearly detected using amateur equipment. Note the lack of variation of the comparison star. Image credit: Arto Oksanen.

horizontal line, you may have a variable star. The photometry package will also provide an error bar (an estimate of how wrong the answer might be) so, by comparing that with the range of the variability seen, you can get a feel for whether the observed variability is real, or maybe just measurement noise.

If you are, instead, looking for fast variations in brightness (such as a rotating asteroid – often with rotation periods of only a few hours), a light curve can again be made but with a series of observations taking place over the course of a single night, rather than weeks. To determine if the observed variability arises from changes in transparency during the night (and star altitude), or is, instead, the real thing, the measured brightness of unvarying comparison stars within the same field of view can also be plotted. If they show a flat curve and the suspected star does not, then the variability is probably real.

In reality, photometry is not quite as simple as it sounds and several problems arise; for example:

- Where does a star image end as you move outward from its bright center? Choosing the correct radius is important.
- Accurate flat-fielding is essential.
- What do you do about photometry in a crowded starfield where the annulus contains a neighboring star?
- How do you measure comets with extended coma?

All these considerations make it more challenging – and more fun – but, in essence, it is quite easy to do and, like astrometry, can be undertaken on a moonlit night when deep sky imaging activities may be impaired.

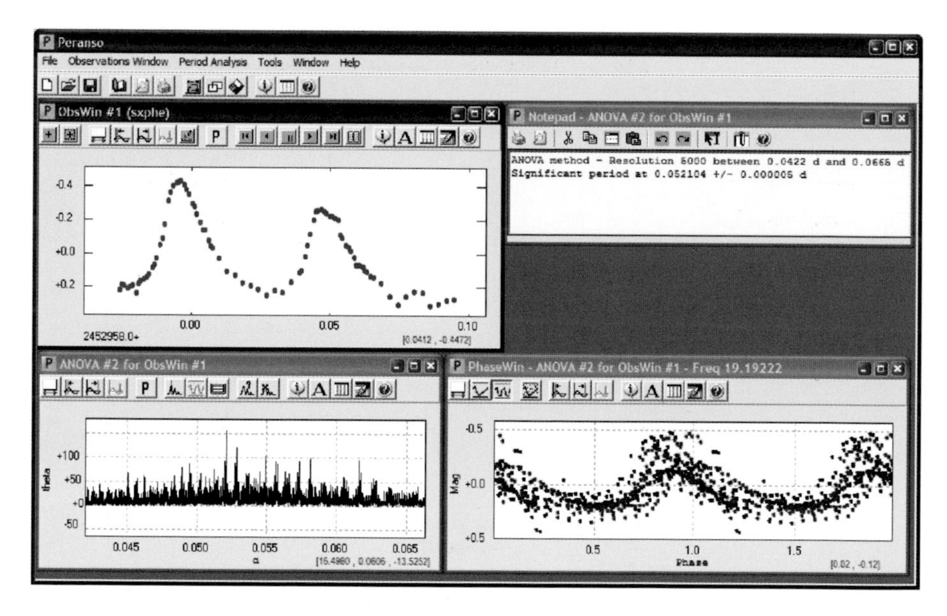

Figure 12.3. *Peranso* from CBA Belgium. A *Windows*-based tool for applying sophisticated analysis techniques to the light curves created using photometry.
Image credit: Tonny Vanmunster.

Packages like *CCDSoft* and *Iris* can be used to extract the photometry results from a series of images. Once that has been done the PC-based *Peranso* or *Unix*-based Starlink package *Period* provide many tools for addressing the subsequent data analysis.

Obviously, astrometry and photometry do not fit in well with what many observers want to get from their observations and so the number of observers pursuing them is relatively small, but, if you ever feel you fancy a break from pursuing the perfect image, these fields represent somewhere interesting to start. The observations that can be undertaken, even with a modest telescope, can be of great value to science, which adds spice to any observing session.

CHAPTER THIRTEEN

The Problems

Image processing is something of a balancing act. It is true that in many cases subtle tweaking of an image can make it look a little better, but there is a point where any extra tweaking brings vanishing returns or, even worse, creates artifacts. Working out when it is time to stop fiddling with an image can be a tricky decision. Stop too soon and you deny yourself the best possible result from your imaging, but keep on going and you, potentially, waste a lot of time you might have used for something more rewarding or productive.

One way to circumvent this problem is to undertake the image reduction and then save the image. Try to spread the image enhancement over several sessions: initially having a preliminary stab at the enhancement using a few favorite techniques, perhaps a contrast stretch, maximum entropy or an unsharp mask. Then store these results and do something else for a while. Coming back to the preliminary results a few days later will make it easier to judge whether or not they truly represented an improvement or were just a waste of time. After a long night of image reduction, or hard day at work, it is very easy to become a little blinkered about the quality of an image and there is a natural tendency to keep tweaking parameters, even though the resultant improvement to the image has become marginal. This is particularly true when the image is *nearly* good enough. Most of us have been guilty of this occasionally. One possible approach is to take the pictures to members of your local society or fellow imagers, ask them to judge which they think are best, and ask for comments. Quite often the flaws within an image will be immediately obvious to others, because they have not sat there staring at it and its precursors for several hours.

As mentioned, the major problem faced by an imager is an image that just isn't good enough in the first place. Possible examples might be an image of Jupiter from a night that had no truly still moments, or an image of the Veil Nebula that had a poor signal-to-noise ratio due to the stacking of too few exposures. In the early days of imaging, it was quite common to look in the magazines and find otherwise attractive CCD images of galaxies that were ruined by elongated field star images or an unsightly sky gradient. But things have moved on a lot since then and the general standard of imaging has improved enormously. You will find that, as time goes by, you will become better at seeing what is wrong with an image, even before you reach the image enhancement stage. Inevitably, you will become better at judging whether an image is likely to reward the considerable effort that may be involved. Most astro-imagers gradually refine their imaging techniques and become more demanding of the equipment they employ.

Consequently, images with poor seeing, a bright sky background or iffy tracking will increasingly be rejected prior to image stacking. All these can be improved but, occasionally, when you are attempting to image something at the limit of the telescope or camera, you will find that the resulting image is just so poor that there is nothing you can do about it with the equipment to hand. At that point median filters, or even gentle smoothing, blurring or binning are sometimes useful techniques to employ, but the image will probably never be something you would want printed on a mug or framed upon your wall. When that happens you have to accept that, while image enhancement can sometimes work wonders, there will be occasions when further effort will be a waste of time.

Outside influences can have an effect upon the images you are dealing with, rendering them difficult to work with. These include the telescope secondary mirror, or camera, dewing up, electrical interference or "pick-up" in the camera or meteors/satellites creating trails across the image.

The art to overcoming these problems is in large part improving your technique by learning how to use dew heaters, untangling electrical cables and taking lots of images among other things. Some things like your neighbors turning on outside lights when they put out the cat, or stray light from passing cars can generally only be overcome with patience – the essential skill for astronomical imaging.

Figure 13.1. Satellite trails – the bane of the long-exposure imager. If you are using short exposures you can median stack this type of unsightly blemish away. Image credit: Ron Arbour.

CHAPTER FOURTEEN

Postscript

So, there we have it, then. We have gone from the basics, into image display through image reduction, image enhancement and the handling of color imagery before finally looking briefly at astrometry and photometry.

Having read all that you might be tempted to imagine we have covered everything, but while what you have read represents a solid foundation, image processing is a huge field of ongoing scientific research. New ideas and ever increasing computer processor power mean that techniques currently under development will eventually become part of the mainstream and will allow us to obtain images even better than those being created at present.

Probably the most important skill needed for successful image processing is one that a book cannot teach – patience.

It is absolutely essential to develop the ability to overcome the temptation to say "that will do", whether it be in relation to telescope focus, in the number of images stacked or the selection of a deconvolution operation. Practice and experimentation will be rewarded. If you rush, you will create many average images, but few that are truly exceptional.

To conclude, the best way of improving your image processing is to just keep on doing it. Practice does not guarantee perfection, but it will mean you have done the best you can – and what more can you ask for?

CHAPTER FIFTEEN

Notes on Image Contributors

The following astro-imagers have generously provided the images that I have used to illustrate this book. Unfortunately, the space available does not permit me to detail each picture shown, nor the equipment used, but looking at their websites will provide much useful information and quickly justify the effort.

Ron Arbour. Widely respected astro-photographer/imager and telescope maker. He has discovered 16 supernovae, an extragalactic nova in M31, three variable stars and one active galaxy. A past Vice-President of the BAA, he was founder of both the Deep sky and Astro-photography sections of the BAA and founder of the UK's Campaign for Dark Skies. He observes with a 300 mm (12 inch) Meade SCT and home-built 406 mm (16 inch) Newtonian from his observatory in Hampshire.

Damian Peach. Widely regarded as one of the worlds' leading planetary imagers, his work has appeared in many books, magazines and webpages. The Assistant Director of the BAA Jupiter and Saturn sections, he has appeared several times on the BBC's *Sky at Night* program, and imaged Mars live for the BBC's *All Night Star Party* in 2003. He was awarded the ALPO's Walter H Haas award in 2004 for outstanding contributions to planetary astronomy. www.damianpeach.com.

Martin Mobberley. A well-known UK amateur astronomer, he images a wide variety of objects. He has written three books for amateur astronomers together with three children's books about astronomy and space travel. He has served as BAA President and received the BAA Goodacre award in 2000. He writes regularly for astronomical magazines and has been a frequent guest on the BBC's *Sky at Night* program. The asteroid 7239 is named after him. uk.geocities.com/martinmobberley

Grant Privett. A member of the Shropshire Astronomical Society, he mainly images from Herefordshire using a home-made TC255-based camera or a Starlight MX716 on a Polaris mounted 250 mm (10 inch) f/4.4 Newtonian. www.shropshire-astro.com/gjp.htm

Nik Szymanek. A keen astronomical photographer based in Essex, he regularly travels to the dark sites of professional observatories to take images. He contributes pictures to astronomical magazines and books, including his own *Infinity Rising*. In 2004 he received the Astronomical Society of the Pacific's Amateur Achievement Award for work in astronomical imaging and public outreach. www.ccdland.net

Gordon Rogers. Observing from the rooftop Crendon Observatory near Oxford he specializes in deep sky imaging using a 406 mm (16 inch) RCOS Ritchey–Chretien, a Paramount ME mount and SBIG cameras. His images are widely used in astronomical publications and he regularly gives talks on imaging deep space. www.gordonrogers.co.uk

Kev Wildgoose. A long-standing committee member of the Shropshire Astronomical Society, he images using a Starlight Xpress MX916 camera for deep sky work and a Philips ToUCam 740k Pro for planetary imaging with his 300 mm (12 inch) Newtonian reflector. www.kevwildgoose.co.uk

CHAPTER SIXTEEN

Appendix

Software

Astronomical image reduction is quite a narrow field, yet despite this, there are a surprising number of software packages and image reduction suites that have been designed for the purpose.

However, the big problem with recommending a particular piece of software is that the development cycle for software is even faster than that for telescopes and camera hardware. It is entirely possible that anything I recommend now may not be the best available on the day you read this book, or will have been supplanted by other packages. However, all of the software I describe below does a good job in its current version and is very unlikely to be less capable in future incarnations.

The best-known software is probably those commercial packages used by *Windows*-based imagers. Of particular note are *Maxim-DL*, *AIP$_4$WIN*, *AstroArt*, *Iris* and *CCDSoft*. Each of these has its own confirmed fans and a considerable number of users. Unless things change greatly, you will find that they differ hugely in price but, while sharing a base level of functionality, some of them are rather more advanced, easy to use and more capable than the others. Before buying any software be sure to look at the websites for each package and keep an eye out for reviews in magazines – beware web reviews; their quality and impartiality is often rather suspect. Better still, before making an expensive purchase, get together with other imagers from your local astronomical society and arrange an opportunity to look over their shoulders while you process some images together. An hour spent using the software with a more experienced user will give you a far better feel for what it can do than reading the packaging or advertising blurb.

Given the rapid growth of the *Linux* operating system in recent years it is well worth mentioning the Starlink Software Collection. Despite the fact it has recently

Figure A.1. *CCDSoft*, a familiar and popular choice among SBIG camera users. Image credit: Gordon Rogers.

been mothballed following a brainstorm at the UK Particle Physics and Astronomy Research Council (who funded it), it will remain available, free, on the web for some years. It provides a whole host of software that can be used for image reductions of professional standard. Packages like *CCDPACK* and *gaia* provide a graphical interface to state-of-the-art algorithms for assembling image mosaics and for image display and enhancement. I am biased on this subject as I wrote some of the Starlink code and can personally vouch for the demanding standards applied, but the fact remains that it is great software, *very* fully documented and free. A possible alternative is the IRAF suite of software packages – which is also free and available for download.

Most of the software mentioned so far has been designed with the astro-imager exclusively in mind. By contrast, the programs *Photoshop* from Adobe and *Paintshop Pro* from Corel are aimed squarely at the normal digital camera user and best suited to the image enhancement phase.

Photoshop is certainly the more powerful of the two, providing pretty much everything you could want (and more besides) and a largely intuitive interface. *Paintshop Pro* is also a very good program and has enough facilities and options to make its instruction manual a book in its own right. The inevitable downside is that the prices of the packages are rather different with *Photoshop* currently costing about three times as much as *Paintshop.*

Some astronomers, especially those creating large mosaics or composite images do virtually all of their image enhancement in *Photoshop,* due to its superb use of

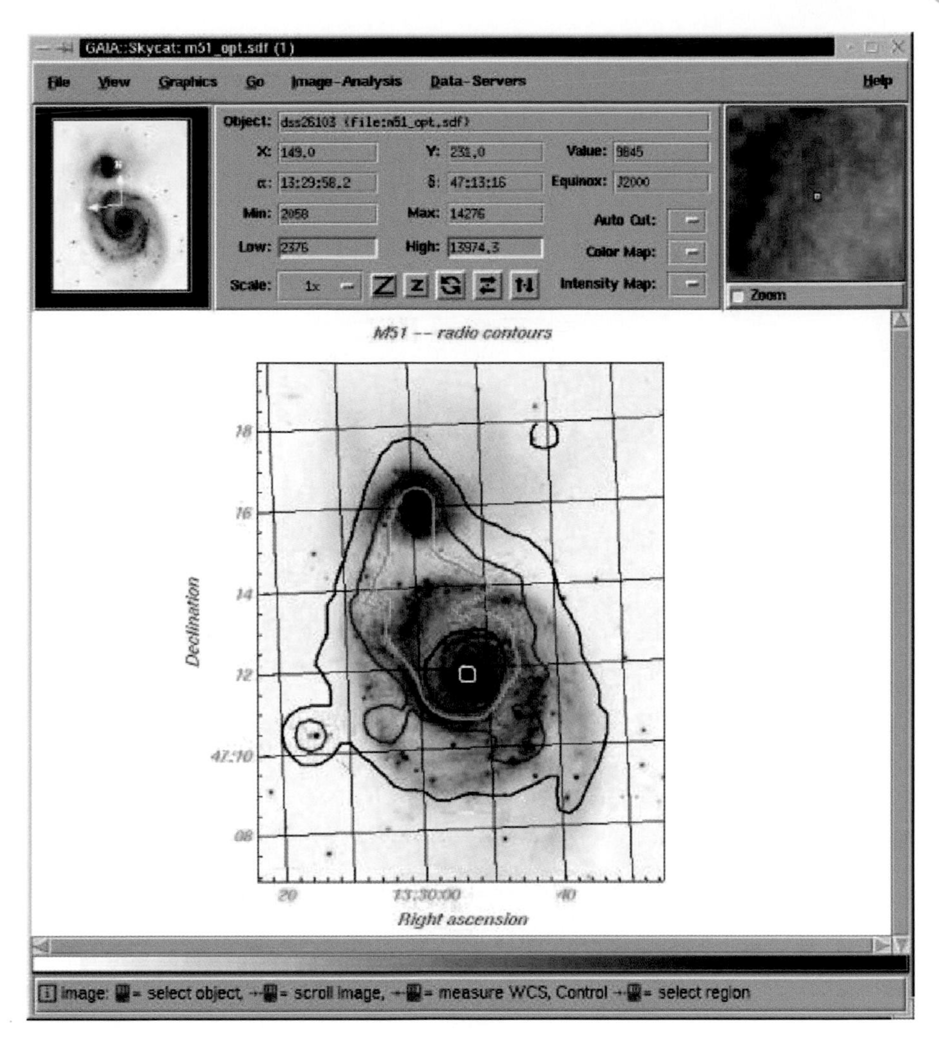

Figure A.2. The *Linux*-based *gaia* software package provides an interface allowing image processing, including astrometry and photometry, to be carried out effortlessly.
Image credit: Peter Draper and Starlink.

layers, the "Curves" function and feathering. Those of us who remember the early years of DOS-based astronomical software can vouch for the fact that the capabilities of the available reduction software have grown enormously with the passing years and that they are gradually acquiring more and more of the features seen in programs like *Photoshop*. So, if the trend continues, it can only be a matter of time before they contain everything the astro-imager needs. Being able to buy just one program fully covering both reduction *and* enhancement would certainly leave more cash free for more exciting things. It is also a great relief to those of us who wrote our own software while waiting for the commercial software to catch up.

Another growing advantage of popular *Windows*-based software packages like *Photoshop* or *AstroArt* is the advent of plug-ins. These are pieces of add-on

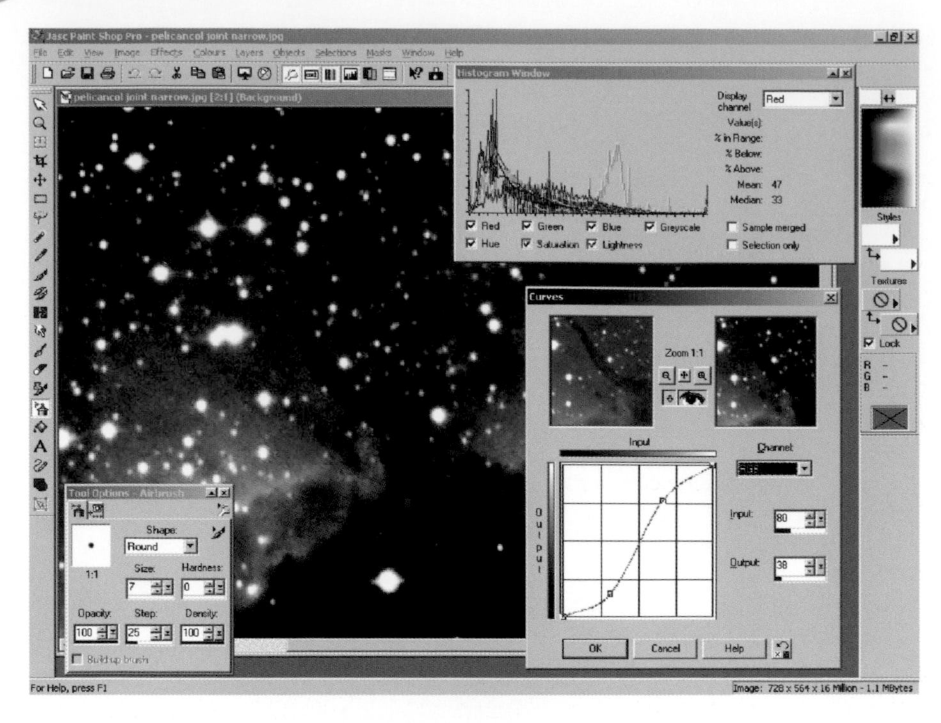

Figure A.3. *Paintshop Pro.* A less expensive, but less powerful alternative to *Photoshop.* Image credit: Grant Privett.

software that extend the abilities by adding more options/functions to their pull-down menus. They are not normally developed by the manufacturer of the main software package, but by astronomers with a penchant for software development, who find they need some functionality the software lacks. For example, *Photoshop* does not import FITS format files, and yet this gap is filled by at least two widely available plug-ins including *FITS Liberator* and *FITSplug*. Similarly, *AstroArt* does not read in imagery created by DSLRS, but there is a free plug-in that does. Plug-ins cover a large range of possibilities. They can sometimes be located directly via the webpages of the manufacturer or alternatively, by a quick search of the internet. Unfortunately, not all are free and few come with any guarantee, but many are very good. Not all packages allow plug-ins as this requires a development effort on the part of the manufacturer, but they are an increasingly common occurrence. Almost a win–win situation for the manufacturer.

On the subject of capturing and processing webcam sequences, it could be said that we are currently spoilt for choice, with several powerful and very sensibly priced candidates to choose from. The best-known processing package is *RegiStax*. There is no denying it is extremely powerful, but it can, for the new user, be a little intimidating in its range of options. Ignoring the software that usually comes with the cameras, the other popular options include *AstroStack*, *K3CCD* and *AstroVideo*. All three are nicely executed and go a long way toward ensuring almost any of us can, with the aid of a webcam, take a good planetary picture. The manner in which they select the best images from an image sequence – whether by wavelet analysis, power spectra or image brightness and statistics – differ quite markedly, but they offer usable default settings that work well enough to get you started. The more experienced user will find much scope for subtle refinements.

Below I list the websites belonging to the main players in the field. There are others, but these are the most commonly encountered.

AIP_4WIN – www.willmanbell.com
AstroArt – www.msb-astroart.com
Astrometrica – www.astrometrica.at
AstroStack – www.astrostack.com
CCDSoft – www.bisque.com
Iris – www.astrosurf.org
K3CCD – www.pk3.org
Maxim-DL – www.cyanogen.com
pcAnywhere – www.symantec.com
Peranso – www.cbabelgium.com
Photoshop – www.adobe.com
Paintshop Pro – www.corel.com
Registar – www.aurigaimaging.com
RegiStax – www.registax.astronomy.net

Hardware

As with software, the hardware specifications are improving rapidly and advance from year to year. But it is fair to say that, over the last few years, few people have done badly by buying instruments from the manufacturers listed below.

DSLRs

Canon Inc. – www.canon.com
Nikon Inc. – www.nikon.co.uk
Pentax – www.pentax.com
Olympus Corporation – www.olympus-global.com

CCDS

Apogee Instruments – www.ccd.com
Starlight Xpress – www.starlight-xpress.co.uk
Santa Barbara Instruments Group – www.sbig.com

Telescopes

Meade Instruments Corporation – www.meade.com
Celestron – www.celestron.com
Orion Optics – www.orionoptics.co.uk
Orion Telescopes and Binoculars – www.telescope.com
Astrophysics – www.astro-physics.com
SkyWatcher/Synta – skywatchertelescope.com

Webcams

ATIK – atik-instruments.com
Lumenera – www.lumenera.com

Philips – www.philips.com
SAC Imaging – www.sac-imaging.com

For the Meade DSI/LPI and Celestron NexImage webcam-like cameras see the URLs given for the appropriate telescope manufacturer.

Further Reading

Any of the books listed below is well worth a read. The list is a long way from being exhaustive, but together with this book they will help to provide you with a really solid understanding of image processing and astro-imaging.

Astronomical Image Processing

CCD Astronomy – Construction and Use of an Astronomical CCD Camera, Christian Buil (Willmann-Bell, 1991).

A fascinating and ground-breaking book with much to recommend it, even 15 years later. Contains some source code and detailed information on the construction of a CCD camera, but don't let that put you off. Just skip those bits.

The Handbook of Astronomical Image Processing, Second Edition, Richard Berry and James Burnell (Willmann-Bell, 2000).

A very impressive achievement, containing a wealth of technical detail, source code and advice, together with the *AIP$_4$Win 2.0* astronomical image processing software package. It seems a little pricey until you notice it comes with the software package.

Handbook of CCD Astronomy, Steve B. Howell (Cambridge University Press, 2000).

An essential guide to image reduction aimed mainly at the research astronomer, but worthwhile for the more adventurous amateur.

The New CCD Astronomy, Ron Wodaski (Multimedia Madness).

A fine book with many useful examples of image processing. Unfortunately, it says little about webcams and DSLRs and focuses quite hard on particular software, but is well illustrated.

A Practical Guide to CCD Astronomy, Patrick Martinez and Alain Klotz (Cambridge University Press, 1998).

Packs a lot of information into quite a thin book. Very much worth a read.

Starlink Cookbooks available from star-www.rl.ac.uk

General Image Processing

The Image Processing Handbook, Second Edition, John C. Russ (CRC Press, 1995).

A well-written book providing a good overall view of the subject. Some of the examples are particularly well chosen and the approach is pretty accessible.

Digital Image Processing, Second Edition, Rafael C. Gonzalez and Richard E. Woods (Prentice Hall, 2002).

A well-respected and comprehensive text on the wider subject. Perhaps a little mathematical for the general reader, but rewarding for those who persist. A third edition should be published in 2006.

Cooled and Long-Exposure Webcams

The QCUIAG website at www.qcuiag.co.uk provides all you could ever want to know about the current state of the art in modified webcams, video surveillance cameras and camcorders. Prominent members have also published several articles in magazines like *Sky & Telescope*. Of particular note was the January 2004 edition, where the Image Gallery section was given over totally to webcam images for a month. It is possible that mainstream manufacturers have now caught up with cameras such as the Meade DSI-Pro II and they may soon render the innovative work of QCUIAG redundant, but it would be best to keep an eye on their website for a year or two yet – just in case.

Acronyms and Abbreviations

CBAT	Central Bureau for Astronomical Telegrams
CCD	charge coupled device– a light-sensitive chip
CD	compact disk – a data storage medium used for computers. Capacity ~700 Mb
CMY	cyan, magenta, yellow – broadband filter colors used in true color imaging
CPU	central processing unit – the heart/brain of a PC
CRT	cathode ray tube – a bulky monitor based upon a vacuum tube
DSLR	digital single lens reflex camera – digital version of an SLR camera
DVD	Digital Versatile (or Video) Disk – the optical data storage medium used for computers. Capacity 2–8 Gb
FT	Fourier transform – a way of converting an image into its component frequencies
FFT	fast Fourier transform – an efficient of way of doing a Fourier transform quickly
FITS	Flexible Image Transport System. The standard format for the exchange of astronomical images
Gb	approximately 1000 million bytes
GUI	graphical user interface – the collection of *windows* used with a mouse to run a computer program
HGSC	Hubble Guide Star Catalogue. A list of 15 million stars (mainly) with reasonably well-known positions and brightness
IC	Index Catalogue of deep sky objects
IR	light with a wavelength in the range 700 nm to 10 microns, i.e. longer than red light
JKT	Jacobus Kapteyn Telescope. A 1 m telescope on La Palma. Part of the Isaac Newton Group
kb	1024 bytes of memory

KBO	Kuiper Belt Object, a planetoid/asteroid orbiting out beyond Neptune
LRGB	luminosity, red, green, blue – three filtered images and one unfiltered for near true color imaging
M	1 million
M	Messier object – deep sky object, e.g. M45 is the Pleiades
Mb	approximately 1 million bytes
NGC	New General Catalogue of deep sky objects
PC	Personal Computer – normally compatible with the *Windows* computer operating system
QE	quantum efficiency – the fraction of incoming light that a light sensor registers
RGB	red, green, blue – broadband filter colors used in true color imaging
SATA	Serial Advanced Technology Attachment – a way of connecting computer disks allowing fast data transfer
SCT	Schmidt–Cassegrain telescope – a compact fork mounted telescope, frequently GOTO
SLR	single lens reflex camera – the type of camera that has interchangeable lenses
SNR	signal-to-noise ratio – the magnitude of signal from a star, etc. divided by the magnitude of the error associated with the measurement. A higher value gives a better image than a low value
TFT	Thin Film Transistor – flat screen LCD monitor
USB	Universal Serial Bus – computer interface standard
UV	ultraviolet light of a wavelength less than 370 nm

Index

Apogee, 17, 137
archiving, 64
AstroArt, 59, 61, 91, 133, 135–137
Astrometrica, 122, 137
astrometry, 16, 51, 77, 88, 121–125, 129, 135
Astrovideo, 115, 136
ATIK, 16, 18, 137
autoguider, 8, 17, 20, 29, 111

batch processing, 48
bias frame, 42, 47–50, 61–62
blurring, 18, 27, 54, 92–96, 98, 110, 128

Canon, 15, 36, 107, 137
Canny, 84
CCDSoft, 61, 125, 133–134, 137
Celestron, 16, 20–21, 28, 137–138
clipped mean, 42, 47, 56
CMY, 14, 106
collimation, 12, 23–24, 30, 119
colour balance, 13, 31, 35, 104, 106–108, 110
condensation, 24–25, 39
curves, 59–60, 107–108, 124, 135

dark frame, 37–42, 44, 46–50, 61–62, 77, 113
dark subtraction, 14, 37, 39, 41, 52, 61, 75
deconvolution, 32, 65, 97–99, 129
deep-sky, 9, 12–13, 15–16, 18, 22, 25, 27, 29–30, 53, 66–67, 95, 106, 108, 110, 124, 131–132, 139–140
DOS, 87, 135
DSLR, 2, 8–9, 15, 18, 22, 30, 32, 39–40, 107, 113–114, 136–137, 139

edge detector, 84–85

FITS, 55, 65, 76–77, 87, 91, 122, 136, 139
flat-field, 14, 44–48, 61, 104
flat-fielding, 11, 14, 44–46, 48, 52, 61, 91, 124
focus, 3, 12–14, 17, 20–24, 32, 47, 53, 79, 92, 104, 119, 129

gaia, 134–135
gradient, 32–33, 38, 46, 49, 59, 62, 84, 88–91, 96, 128

H-alpha, 32, 108–110

IRAF, 136
Iris, 125, 133, 137

JPG, 65
Jupiter, 7, 9, 29, 99, 118, 127, 131

K3CCD, 136–137

Larson-Sekanina, 32–33
Linux, 133, 135
Lucy-Weiner, 97
Lumenera, 8, 28, 137

Mars, 9, 27, 29, 105–106, 118, 131
Maxim DL, 59, 113–114, 116, 133, 137
maximum entropy, 65, 97–98, 127
Meade, 18, 20, 28, 31–32, 87, 131, 137–139
mean, 42, 47, 50, 54, 56, 62, 75, 78, 94
median, 42, 47, 54, 61–62, 75
median filter, 93–95, 106, 110, 113, 128
median stack, 55–57, 60, 93, 95, 128
mode, 15, 36, 65, 75, 79, 105
moon, 1, 4, 7, 16, 20, 27–29, 59, 81, 88–90, 95–96, 106–107
mosaic, 56–60, 134

N-II, 108
narrow-band filter, 32, 108–111
Nikon, 15, 137

O-III, 108–109
occulting strips, 22
Olympus, 137
Orion (US), 39–40, 95, 108
Orion (UK), 39–40, 137

Paintshop Pro, 60, 134, 136, 137
Pentax, 15, 137
Peranso, 124–125, 137
periodic error, 20, 77
pixel, 3–9, 13–15, 22, 29, 36, 38–39, 42, 44–52, 54–55, 57–60, 62–63, 65–67, 69–71, 75–81, 84, 92–97, 101–102, 105, 113, 117–118, 121–123
Philips, 8, 16, 28, 132, 138
photometry, 16, 51, 54–55, 77, 121–125, 129, 135
Photoshop, 59, 86, 91, 116, 134–137
plugin, 91, 136
power spectrum, 116, 136

QCUIAG, 40, 139

radial gradient, 32
Registar, 59, 137
RegiStax, 14, 28, 115–116, 119, 136–137
resolution, 4–8, 13, 15, 23, 31–32, 35–36, 51, 59, 105, 113, 117–118
RGB, 14, 29, 54, 102, 104–109, 111, 118, 140
rotation, 16, 50, 54, 124

S-II, 108–109
SAC, 13, 16, 18, 39, 138
sampling, 7, 11, 89–90
Saturn, 7, 9, 11, 28–29, 98, 105–106, 117–118, 131
SBIG, 2, 4, 6, 9, 13, 16–17, 50, 87, 132, 134, 137
scaling, 25, 35–36, 50–51, 57, 66–71, 73, 84, 88, 96, 107–108
SCT, 11–12, 16, 20–21, 25, 30, 54, 105, 131, 140
Sobel, 84
Starlight Xpress, 6, 9–10, 17, 50, 102, 132, 137
Starlink, 125, 133–135, 138
statistics, 50, 75–76, 136–137

transfer function, 60, 68, 70–71, 73
transformation, 50–51, 54
translation, 50–51, 54

unsharp mask, 32–33, 65, 84, 91, 95–97, 127
USB, 2, 9–10, 87, 114, 140

wavelets, 116–118, 136

Printed in Singapore